Housing Policy, Wellbeing and Social Development in Asia

This book investigates how housing policy changes in Asia since the late 1990s have impacted on housing affordability, security, livability, culture and social development.

Using case study examples from countries/cities including China, Hong Kong, India, Japan, Taiwan, Korea, Malaysia, Bangladesh, Singapore, Indonesia, Thailand and Vietnam, the contributors contextualize housing policy development in terms of both global and local socio-economic and political changes. They then investigate how policy changes have shaped and re-shaped the housing wellbeing of the local people and the social development within these places, which they argue should constitute the core purpose of housing policy.

This book will open up a new dimension for understanding housing and social development in Asia and a new conceptual perspective with which to examine housing which, by nature, is culture-sensitive and people-oriented. It will be of interest to students, scholars and professionals in the areas of housing studies, urban and social development and the public and social policy of Asia.

Rebecca L. H. Chiu is Professor and Head of the Department of Urban Planning and Design and Director of the Centre of Urban Studies and Urban Planning at the University of Hong Kong, Hong Kong SAR, and Founding Chairman of Asia Pacific Network for Housing Research.

Seong-Kyu Ha is Emeritus Professor, Department of Urban Planning and Real Estate, Chung-Ang University, Seoul, Korea and President of the Korea Research Institute of Housing Management, Seoul, Korea.

Routledge Studies in International Real Estate

The Routledge Studies in International Real Estate series presents a forum for the presentation of academic research into international real estate issues. Books in the series are broad in their conceptual scope and reflect an inter-disciplinary approach to Real Estate as an academic discipline.

Oiling the Urban Economy
Land, Labour, Capital, and the State in Sekondi-Takoradi, Ghana
Franklin Obeng-Odoom

Real Estate, Construction and Economic Development in Emerging Market Economies
Edited by Raymond T. Abdulai, Franklin Obeng-Odoom, Edward Ochieng and Vida Maliene

Econometric Analyses of International Housing Markets
Rita Li and Kwong Wing Chau

Sustainable Communities and Urban Housing
A Comparative European Perspective
Montserrat Pareja Eastaway and Nessa Winston

Regulating Information Asymmetry in the Residential Real Estate Market
The Hong Kong Experience
Devin Lin

Delhi's Changing Built Environment
Piyush Tiwari and Jyoti Rao

Housing Policy, Wellbeing and Social Development in Asia
Edited by Rebecca L. H. Chiu and Seong-Kyu Ha

Housing Policy, Wellbeing and Social Development in Asia

Edited by Rebecca L. H. Chiu and
Seong-Kyu Ha

LONDON AND NEW YORK

First published 2018
by Routledge
2 Park Square, Milton Park, Abingdon, Oxon OX14 4RN

and by Routledge
711 Third Avenue, New York, NY 10017

Routledge is an imprint of the Taylor & Francis Group, an informa business

British Library Cataloguing-in-Publication Data
A catalogue record for this book is available from the British Library

Library of Congress Cataloging-in-Publication Data
Names: Chiu, Rebecca Lai-Har, editor. | Ha, Sæong-gyu, 1947- editor.
Title: Housing policy, wellbeing and social development in Asia / edited by Rebecca L.H. Chiu and Seong-Kyu Ha.
Description: New York : Routledge, 2018. | Series: Routledge studies in international real estate | Includes bibliographical references and index.
Identifiers: LCCN 2017058515 | ISBN 9781138208186 (hardback ; alk. paper) | ISBN 9781315460055 (ebook ; alk. paper)
Subjects: LCSH: Housing policy–Asia.
Classification: LCC HD7358.A3 H6825 2018 | DDC 363.5/561–dc23
LC record available at https://lccn.loc.gov/2017058515

ISBN: 978-1-138-20818-6 (hbk)
ISBN: 978-1-315-46005-5 (ebk)

Typeset in Goudy
by Wearset Ltd, Boldon, Tyne and Wear

Contents

Figures

Tables

Contributors

Ainoriza Mohd Aini is a Senior Lecturer and is currently the Programme Coordinator for the Master's in Real Estate at the University of Malaya. She completed her PhD in the areas of sustainability and responsible property investment. Her current research interest is financing issues for first-time homebuyers and housing challenges for migrants. In the past she has completed national studies on elderly housing, the Housing-Income Index© and corporate real estate as well as sustainability. Dr Ainoriza has been recently been selected as the Youth Member in the TN50 Circles of the Future (Living & Well-being). Her role is to collaborate with other youth leaders and experts to develop recommendations for TN50.

Wan Nor Azriyati Wan Abd Aziz is an Associate Professor in the Department of Estate Management on the Faculty of Built Environment at the University of Malaya, Kuala Lumpur. She holds a doctorate in Housing Policy from the University of Dundee, Scotland, UK. Her areas of expertise and research interests include housing policy and issues, urban studies and property development. She has published many articles in international journals and presented at conferences and seminars related to housing and land development. She has also vast experience in conducting research and consultation projects commissioned by the state and federal government related to housing. Wan Nor Azriyati is currently Deputy Dean of Sustainability Science Research Cluster. She was formerly the Head of the Department of Estate Management on the Faculty of Built Environment. At the national level, she is a Board Member, Board of Valuers, Appraisers and Estate Agents, Malaysia; at the international level, she is appointed as a Governing Council member of the ASEAN Valuers Association.

Chin-Oh Chang is a Distinguished Professor in the Department of Land Economics and the Director of the Taiwan Real Estate Research Centre, National Chengchi University, Taipei, Taiwan. He received his Master's in Architecture from MIT, USA in 1980, and a City and Regional Planning Ph.D. at the University of Pennsylvania, USA in 1986. He was President of the Asian Real Estate Society (AsRES) in 1997–1998, and President of the Global Chinese Real Estate Congress (GCREC) in 2009–2010. His research is published in *Urban Studies*, *Housing Studies*, *Journal of Property Research*, *Habitat International*, *International Real Estate Review* and several other journals. He has concentrated on areas related to housing and land policy, real estate investment and financial analysis.

Jie Chen is a Professor of Real Estate at Shanghai University of Finance and Economics (SHUFE). His research covers various fields in regional, urban and housing-related economic issues. He has published more than 30 papers in reputed international journals and also authored or edited five books including *The Future of Public Housing: Ongoing Trends in the East and the West (2013)*. He works as a housing policy advisor for Chinese central government agencies, as well as a senior consulting expert for a number of international organizations. He is on the Board of Trustees of the Urban Studies Foundation, on the editorial board of *Housing Studies*, a board member of the Asian Real Estate Society and a steering member of the Asia Pacific Network for Housing Research.

Rebecca L. H. Chiu is a Professor and Head of the Department of Urban Planning and Design and Director of the Centre of Urban Studies and Urban Planning at the University of Hong Kong. Her current research interests include housing and urban sustainability issues in high-density Asian cities, especially in China, comparative housing policies in Asia and housing policy transfer. She is the Founder Chairman of the Asia Pacific Network for Housing Research. She has been appointed to government boards and appeal panels related to housing, planning, urban renewal and natural and heritage conservation in Hong Kong.

Sadeque Md Zaber Chowdhury works for the Bangladesh Government and currently holds the position of Additional Superintendent of Police. He is responsible for the overall physical infrastructure planning and development of the Bangladesh Police. He received his PhD from The University of Hong Kong. His research interests involve Housing Policy Studies, Planning Regulatory and Institutional Regimes and Urban Environmental Crime. He is also a practicing urban planner and a member of the Bangladesh Institute of Planners.

Chua Beng Huat is a Professor in the Department of Sociology, National University of Singapore. His research interests include housing and urban studies, comparative political economy of Southeast Asia and cultural studies in Asia. His most recent book is *Liberalism Disavowed: Communitarianism and State Capitalism in Singapore* (2017) and he is guest editor of a special issue, "Inter-referencing East Asian Occupy Movements", of the *International Journal of Cultural Studies* (2017). He is a founding co-executive editor of the *Inter-Asia Cultural Studies Journal*.

Seong-Kyu Ha is an Emeritus Professor of Urban Planning and Real Estate at the Chung-Ang University, South Korea. He also serves the president of the

Korea Research Institute of Housing Management. He received his PhD from the University College London and an MSc in Urban Planning from the London School of Economics. His current research interests have centred on low-income housing policies, urban residential regeneration and housing management. He has authored numerous publications on housing policy and urban and community regeneration, including *Housing Policy and Practice in Asia* (Croom Helm, 1987). Recently he was elected Chairman of the Korea Housing Service Society, a housing expert consulting and research group in Korea.

Noor Rosly Hanif is an Associate Professor in the Department of Estate Management on the Faculty of Built Environment at the University of Malaya (UM), Kuala Lumpur. He was the Dean of the Faculty of Built Environment, UM, from 2009 to 2014. He is also a professional registered Valuer and Estate Agent with The Board of Valuers, Appraisers & Estate Agents Malaysia, Fellow Member of the Royal Institution of Surveyors Malaysia (FRISM) Malaysia and Member of the Royal Institution of Chartered Surveyors (RICS) United Kingdom. He is as an External Examiner of University Tunku Abdul Rahman (UTAR) and on the Board of Academic Governance INSPEN Ministry of Finance Malaysia. Dr Hanif is also an expert panellist for the Malaysian Qualifications Agency (MQA) to accredit new programs in Real Estate in Malaysia since 2010. His ongoing research is in affordable housing, *waqf* development, urban heat island, ageing shelter needs and auctioneering.

Yosuke Hirayama is a Professor of Housing and Urban Studies at the Graduate School of Human Development and Environment at Kobe University, Japan, working extensively in the areas of housing and urban change, homeownership and social inequalities, as well as comparative housing policy. His work has appeared in numerous international and Japanese academic journals, and he is a co-author of *Housing in Post-Growth Society: Japan on the Edge of Social Transition* (Routledge) and also a co-editor of *Housing and Social Transition in Japan* (Routledge). He has received academic prizes from the City Planning Institute of Japan, Architectural Institute of Japan and Tokyo Institute of Municipal Research. He is also a founding member of the Asia Pacific Network for Housing Research.

Bor-Ming Hsieh is an Associate Professor and Head of the Department of Land Management and Development and Head of the Bachelor's Program in Real Estate Finance at Chang Jung Christian University, Tainan, Taiwan. He received his Ph.D. in Urban Studies at the University of Glasgow, UK, in 2002. His current research interests centres on housing market, spatial analysis of housing prices, housing policy and urban regeneration. His research is published in the *Journal of Real Estate Research, Journal of Housing Studies, City and Planning, Journal of Statistics and Management Systems* and several other journals. He is on the Board of Directors of the Asian Real Estate

Society and the Steering Committee Member of the Asia Pacific Network for Housing Research.

Misa Izuhara is a Reader in Comparative Policy Research at the School for Policy Studies, the University of Bristol, UK. Misa has been undertaking extensive research internationally in the areas of housing and social change, the life-course and intergenerational relations and comparative policy analysis. Her research projects include cross-national comparative research on 'Housing assets and intergenerational dynamics in East Asian societies' funded by the UK Economic and Social Research Council. She is the author of *Housing in Post-Growth Society: Japan on the Edge of Social Transition* (with Yosuke Hirayama, Routledge 2018) and *Housing, Care and Inheritance* (Routledge, 2009) and the editor of *Handbook on East Asian Social Policy* (Edward Elgar, 2013). She is currently the Co-Editor of the peer-reviewed international journal, *Journal of Social Policy*.

Mandy H. M. Lau is an Assistant Professor in the Department of Urban Planning and Design at the University of Hong Kong. She received her PhD and MPhil from the Department of Land Economy at the University of Cambridge and a BSc in Sociology from the London School of Economics. She is currently Secretary of the Asia Pacific Network for Housing Research. Her primary research interests include urban governance, contentious urban developments, inadequate housing, and planning for affordable housing. Recent publications include: "Framing Processes in Planning Disputes: Analysing Dynamics of Contention in a Housing Project in Hong Kong" in *Housing Studies* (2017) and "Tackling Uncertainties in Plan Implementation: Lessons from a Growth Area in England" in *Town Planning Review* (2015).

Thammarat Marohabutr is a Lecturer in the Department of Society and Health of Mahidol University. His research interests lie in the fields of social policy and political economy, including housing and health policy, civil movements and welfare administration. He contributed an article, "Bangkok's Housing Market and Its Trends: A Slowdown from Recovery since 1997 Economic Crisis", to *Housing Express* (Hong Kong: Chartered Institute of Housing, 2008), and completed a PhD thesis, "Housing Policy in Thailand: Implications for Welfare Typology" (2011), at the City University of Hong Kong. He is also a member of the Network of East Asian Think Tanks (NEAT) Working Group on Inclusive Growth.

Urmi Sengupta is an architect/town planner and is currently affiliated with the School of Natural and Built Environment, Queen's University Belfast, UK. Urmi's research interest lies at the intersection of housing policies, practice and urban transformation, especially policies that seek to address poverty and inequalities in cities of the global South. Her recent research has focused on the trends and issues in housing in Asia, housing polices and markets, public space and post-disaster space and reconstruction. She has edited a book, *Coming of an Age: Trends and Issues in Housing in Asia*, 2017 (Routledge with

A Shaw) and is currently working on her new book, *Urbanism in the Global South: Public Space in Cities in Transition*.

Bokyong Seo is a Research Associate in the Department of Urban Planning and Design at the University of Hong Kong. She received her Ph.D. in Urban Planning from the University of Hong Kong and her MSc in Urbanism from the Delft University of Technology. Her research interests have been focused on housing policy and social changes, urban shrinkage, urban regeneration and gentrification and age-friendly cities. Her major publications include: 'Dual Policy to Fight Urban Shrinkage: Daegu, South Korea' in *Cities* (2017); and 'Social Cohesiveness of Disadvantaged Communities in Urban South Korea: The Impact of the Physical Environment' in *Housing Studies* (2014).

Connie Susilawati is a Course Coordinator and Senior Lecturer in Property Economics at Queensland University of Technology, Australia. Her passion for providing real-world learning and international opportunities shows in her teaching, research, leadership and engagement. Her work focuses on housing policy for both developing and developed countries. She led the Economic Impact on Sustainable Housing section of an ARC-Linkage funded project. Susilawati offers multiple international education opportunities for both Australian and Indonesian students/leaders in infrastructure, asset management and property development. She has worked for the largest housing development company (Ciputra Group) and led property students to attend summer program at the University of Surabaya, Indonesia.

Hoai Anh Tran is an Associate Professor in Urban Studies at Malmö University, Sweden, and the Program Manager of the graduate program in Urban Development and Planning at the Department of Urban Studies, Malmö University. She researches urban development and housing policies, urban planning and socio-spatial consequences, modernization and urban changes with a focus on Vietnam. Her most recent project deals with urban space production and urban qualities with examples from the new urban areas of Hanoi.

Meisen Wong is a PhD Fellow in the International Graduate Research Program, Centre for Metropolitan Studies (CMS), Berlin Technical University. Her current research is on 'ghost cities' in China.

Mahazril 'Aini Yaacob is currently a Lecturer on the Faculty of Administrative Science and Policy Studies at Universiti Teknologi MARA, Seremban Campus. She is also a PhD student in the Department of Social Administration and Justice, Faculty of Arts and Social Sciences, at the University of Malaya. Her research interests focus on social policy, housing studies, public administration, public management and human resource management. She has also experience in conducting research and has acted as principal investigator in several research projects. She has experience teaching subjects related to Human Resource Management, Industrial Relations, Organisational Behaviour, Urban Sociology and Public Administration.

Ngai Ming Yip is a Professor in the Department of Public Policy at the City University of Hong Kong and convenor of the Urban Research Group in the department. He researches housing and urban policy and governance of the neighbourhood as well as urban activism in East and Southeast Asian countries. He publishes extensively in international journals. He is an editor of *Housing Studies* and participates actively in the professional and policy communities. He is a member of the commercial property committee of the Hong Kong Housing Authority and chairman of the Chartered Institute of Housing Asian Pacific Branch (2013, 2014).

Preface

This book has been gestating for some years. Since the establishment of the Asia Pacific Network for Housing Research (APNHR) at The University of Hong Kong in 2001, bi-annual conferences have been held for housing and related researchers working on Asia Pacific and beyond to meet and exchange ideas. In the meantime, while book volumes on housing in Asia were published, none were led by Asian housing researchers themselves or included a more comprehensive social dimension. Thus, the idea of filling the gap came, and in some ways it is incumbent upon APNHR to take the lead as it has members and connections throughout the Asian region. While outsiders' views and international perspectives are as important as those of the insiders, perhaps the more intricate social dimensions of housing are better experienced and told by local researchers, especially if the understanding is framed by conceptual interpretation and compared with international experiences. There nonetheless still need to be platforms for local researchers to disseminate their insights and visions. This book provides such a platform, not only discussing over a dozen Asian countries but also attempting to include all major housing systems of various stages of development and socio-economic and political circumstances.

The aim of the book is thus to investigate how social changes have impacted housing policy in Asia since the late 1990s and, in turn, how housing policy changes have impacted housing wellbeing and social development, including the welfare implications of housing policy. It seeks to answer the following research questions collectively and in each individual housing system where appropriate:

1 How have global economic trends and local socio-economic and political conditions incurred housing policy changes since the late 1990s, given the path-dependent nature of housing and the role of housing in addressing the local socio-economic and political challenges?
2 What are the effects of housing policy changes in enhancing housing wellbeing, in terms of housing affordability, housing security, livability and housing culture?
3 Have the policy changes transformed the welfare nature of housing policy and their social functions, and if so, how?

The chapters are organized according to geographical locations, starting with East Asia, and then South East Asia and South Asia. Given the diversity in economic conditions, social environment, housing culture and urban contexts, the focuses of discussion vary; however, all attempt to address the above issues from specific dimensions and critical lenses. Although naturally we could not include all Asian countries, we are confident that this volume provides a much-needed insiders' interpretation of housing policy change and social development in rapidly developing Asia, and hopefully will stimulate further research on this important topic.

To accomplish this book project, we are indebted to Elizabeth M. Fox for the text-editing, Sandra Mather for re-drawing the figures, and Bokyong Seo for co-ordinating the editorial work. Most importantly, we would like to thank all contributors for sharing their fascinating housing stories. We believe that our efforts give the world insights into our housing experience. We await feedback from near and far to enable better housing and social development in the future Asia.

Rebecca Lai Har Chiu and Seong-Kyu Ha, Editors
November 2017

1 Introduction

Conceptual contexts

Rebecca L. H. Chiu

Preamble

The few book volumes on housing in Asia have discussed mainly the socio-economic and political factors and functions of housing policy. None has comprehensively investigated the impact and causes of housing policy changes on housing wellbeing and social wellbeing, especially those brought about by the sharp economic fluctuations, social trends and rapid urbanization since the late 1990s. The book of Groves, Murie & Watson (2007) aims to investigate the role of housing in welfare states and the emerging importance of asset-based welfare in six Asian countries and compares them generally to those of the U.K. and European countries. They argue that expanding ownership rather than citizens' right marks a significant departure of the Asian models from the European ones, although the latter has actually reversed the trends to be more akin to the Asian cases. The welfare function of housing policy is undoubtedly an important topic, although it is only one social dimension of housing and social wellbeing. The book of Doling & Ronald (2014) is concerned with how the housing systems in East Asia have contributed to the success of the region, as well as how they are adapting to new, more challenging conditions and how they are likely to fare in the future. Thus, although it covers nine Asian countries, its focus of investigation is to identify and update the East Asian housing model in its varied economic, demographic and developmental contexts. This book, in contrast, addresses the housing wellbeing and social outcomes of the housing policy changes triggered by social and broader transformations in twelve different Asian cultural settings.

While this book contextualizes Asia's housing policy development in the global and local socio-economic and political changes and the non-housing functions of housing policy, its central theme is how social changes have triggered housing policy changes, which in turn have shaped and re-shaped the housing wellbeing of the local people and the social development of a place. Housing as an element of social development encompasses the security of basic living standards not only in the short-term but also after retirement, a safe living environment and the community development associated with the place of residence. Underpinned by these objectives, this book attempts to open up a new dimension for understanding the relationships between and among housing, social change and social development in Asia and to innovate a new conceptual perspective to understand housing, which, by nature, is culture-sensitive, people-focused and socially-constituted. We thus need to form a conceptual foundation for this study.

Concepts

Although this book contextualizes the housing policy development of Asia in the global and local socio-economic and political changes and the non-housing functions of housing policy, as described above, its central theme is how the different social settings and their changes have influenced housing policy development which subsequently shaped and re-shaped the housing wellbeing of the local people specifically and the social development of a place generally. We believe that improving housing and social wellbeing constitute a core purpose of housing policy. *Housing wellbeing* encompasses affordability, security and the tangible and intangible aspects of liveability, qualified by the specific housing culture of a locality, for example, acceptance of housing space standards, residential form and the social meaning of housing. *Social development related to housing policy* pertains to the security of basic living standards not only in the short-term but also in the long-term (thus the welfare nature of housing policy is relevant), a safe living environment, community development in terms of social cohesiveness reflected in the sense of community, identity, neighbourliness and social inclusiveness, as well as the ability of housing to enhance other forms of the wellbeing of its users such as the post-retirement protective function of home ownership (Chiu, 2004; Grzeskowiak, Sirgy, Lee & Claiborne, 2006; Hulchansky, 1995; Seo & Chiu, 2014).

This book interprets *housing policy* as a form of organized effort designed and orchestrated by the government to make use of public and private resources and both regulatory and other means to solve housing problems to improve housing standards as well as to bring about the wellbeing of a society through the housing endeavours. The policy approaches, or forms of government intervention in the housing sector, are diversified and influenced by political and economic ideology, resources availability, governance modes and specific policy objectives. Accordingly, the extent and nature of government subsidies form a spectrum, spanning a range from in-kind, through regulatory means, to in-cash and hybrid forms. It needs to be acknowledged, nonetheless, that in some places, policy goals and policy tools may not be formulated or available, or even if they are, enforcement may not be forthcoming (Chiu, 2008).

The meaning of *social development* varies in different contexts and to different people. Midgley (2014) summaries its attributes: it involves processes; it is progressive in nature; it is part of a larger multifaceted process; it is interventionist as it requires human agency to take action; it is productivist as it contributes to economic development; it is universalistic in scope involving the whole community; it is committed to promote social wellbeing. Pawar & Cox (2010) summarize the definitions into three categories. First are definitions that emphasize the need for systematic planning and the inseparable relationship between social and economic development. These definitions take social development as a totality comprising economic, political, social and cultural aspects, and the aim of development is to improve people's general welfare. Second are definitions that focus on producing structural change as the core element of developing the society and the change involves the re-organization and reorientation of entire economic and social systems. Third are definitions that focus on achieving human potential, fulfilling needs and attaining a satisfactory quality of life. Given the multifaceted nature of housing and its intricate relationships with all sectors in a society, this book adopts all three categories of definitions. The first two categories are pertinent to the diversity of macro social changes and growth that Asian countries are going through, and the third are specific social wellbeing that housing policy intend to produce. Social development in this volume therefore refers to the whole societal contexts and more fundamental structural reforms that precipitate housing policy changes as discussed above and improve the quality of life; more equitable distribution of resources; greater public participation in making decisions on housing affairs; inclusion and empowerment of disadvantaged groups; improvement of the relationships between people and social/economic institutions and those between human needs and social policies and programmes; and release of human potential to eliminate social inequities and problems and to enhance life-sustenance, self-esteem and freedom and to improve relationships between people and their institutions.

The definitions of *social wellbeing concepts* vary across studies, and each chapter defines the terms in its own context. The following indicative definitions were supplied to the book contributors for reference at the outset. Social cohesion: a state of interactions among members of society characterized by trust, a sense of community and the willingness to participate and help, as well as the actual supportive behaviour (modified from Chiu, 2004 and Chan, To & Chan, 2006). Sense of community: feeling and acknowledgement of being part of a larger, dependable and stable social structure, reflected in emotional attachment, shared identity and frequent social interaction (Chiu, 2004; Unger & Wandersman, 1985). Social capital: a set of productive resources underlying interpersonal trust, trustworthiness, solidarity, reciprocity and engagement in community affairs (Chan et al., 2006). Social inclusion: a situation where there is universal access to resources and decision-making processes and mutual respect for differences and ability to contribute (Cappo, 2002; Oxoby, 2009).

Chapter themes: housing stories of Asia

We organize the chapters according to geographical locations, starting with East Asia, and then South East Asia and South Asia (Figure 1.1).

For the East Asian region, the chapter on China contextualises and interprets the transformation of housing provision regimes in urban China since the

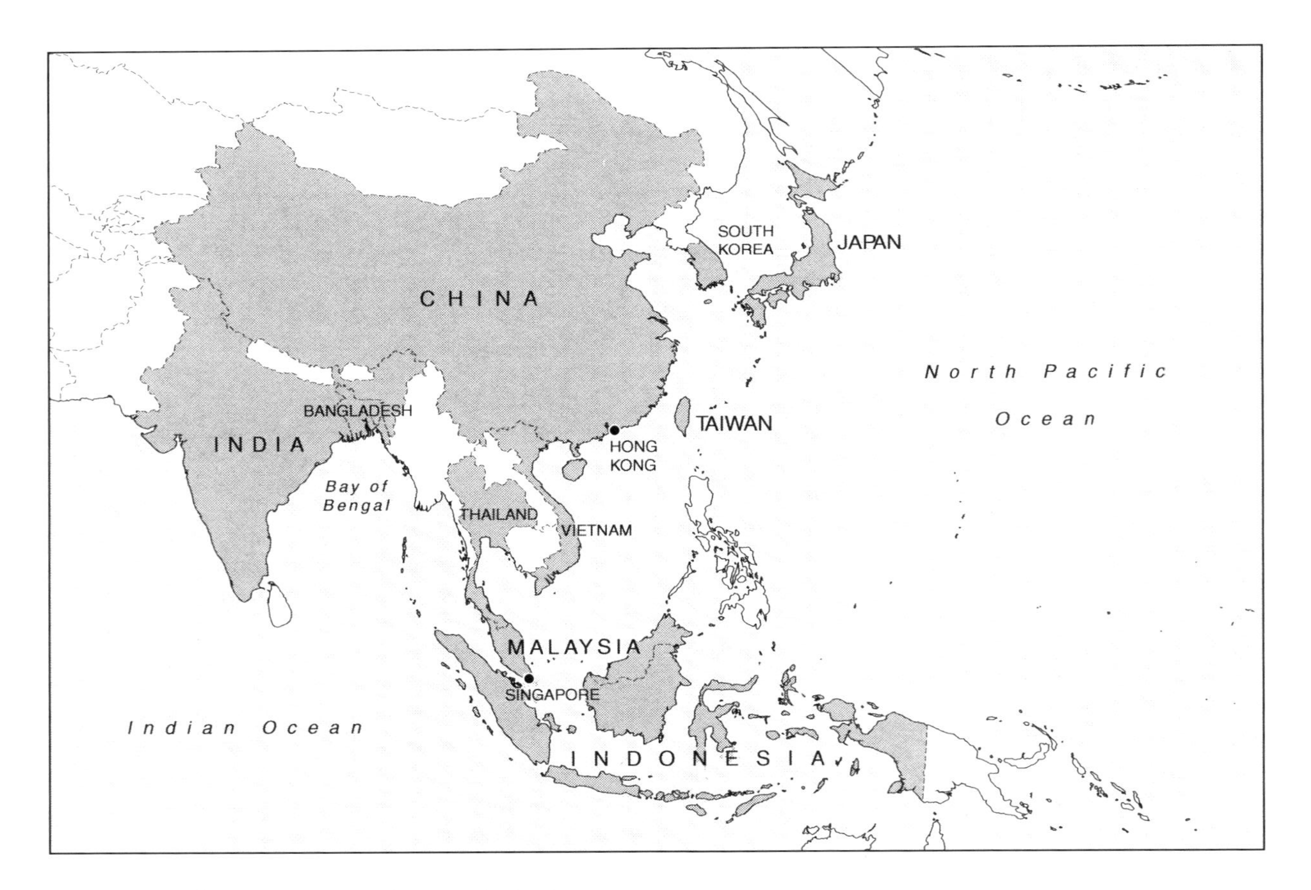

Figure 1.1 Twelve Asian countries/cities in the book.

abolishment of the urban welfare housing system in 1998. Public rental housing provision in Shanghai is examined to offer insights into the complexity of the societal-economic dynamics in post-reform urban China. The chapter on Hong Kong elucidates the social changes that have given rise to the dominance and limitations of public rental housing and their implications and impacts on the social wellbeing of the lower income groups. This chapter argues that the rental-biased subsidy policy constitutes a security-based welfare policy engendering welfare benefits similar to that offered by an asset-based welfare policy in terms of housing security, affordability and social wellbeing. The Japan chapter focuses on the changing role of housing in Japan's transitions into a post-growth and super-aged society, exploring the key issues of shrinking housing and mortgage markets, affordability problems in the deflationary economy and housing market and generational fractures associated with housing. The Korea chapter acknowledges the vast improvement in housing supply and government intervention in housing markets, warns of the persistent divisiveness of housing issues to the country and calls for the government's attention to the social function of housing, especially its implications for the vulnerable groups. The Taiwan chapter points out that housing policy, housing wellbeing and housing justice have dramatically changed in the past several decades, the most significant being re-focusing the goal of housing policy on enabling every citizen to live in an appropriate home. However, the skyrocketing house prices and the low holding costs for property owners have worsened housing inequality and injustice.

In the South East Asian region, the Singapore chapter points out that as the economy enters into slow growth with inflation in costs of living and housing assets, groups excluded from home ownership appear, even with government subsidies. As home ownership is linked to retirement funding, the government is caught in the dilemma of being unable either to take radical measures to deflate housing prices or to expand rental schemes as they reinforce rising housing inequality. The Malaysia chapter demonstrates that the government plays an important role to ensure adequate access to home ownership by all. The role of the state has evolved from enabling only the bottom 40 per cent income groups to home ownership to include the middle-income households.

In contrast to the East and South East Asian countries and cities, the South Asian countries rely more heavily on the informal sector to solve the housing problems of the lower-income families. The India chapter traces the shifts and continuities in housing policy development in the past decades, contending that India's housing transformation remains unfinished. The hurdles are little or no penetration of housing finance into the low-income segments, weak institutional structure and resource base of the implementing bodies and governance gaps in the broad (non)coalition of interests across the community spectrum. The Bangladesh chapter argues that the formal housing market in Dhaka fails to provide affordable housing to the middle-income groups, who struggle to maintain a minimum standard of living and to access affordable rental housing. The regulatory and infrastructure development regimes and the institutional environments are found to be non-enabling, constituting the major underlying causes of the severe housing affordability problems faced by the middle-income groups, who are the pillars of the economy. The Vietnam chapter elucidates the huge increase in housing inequality consequent to the housing reform. It highlights the complex state-market relation that leads to the paradox of the state-supported, corporate-led formal sector's producing a small share of urban housing while the much-suppressed informal sector supplies the majority of urban housing. The Thailand chapter delineates the big divide in housing conditions between the middle- to high-income segment and the low-income segment and calls for attention to this blatant inequality. The slum upgrading scheme advocating participatory democracy is seen as a way to push forward the living standards of the urban poor. Finally, the Indonesia chapter considers the diverse approaches adopted to enable different income groups to acquire home ownership. Housing finance schemes, subsidized housing improvement programs and private housing with government subsidies and support are provided to different income groups. However, the lack of affordable housing in the city centre has increased transportation cost and time and worsened the social well-being of vulnerable groups.

Acknowledgement

The chapter described in this chapter was funded by grants from the Research Grants Council of the Hong Kong Special Administrative Region, China (Project no. HKU 742811H).

References

Cappo, D. (2002). Social inclusion initiative: Social inclusion, participation and empowerment. *Address to Australian Council of Social Services National Congress*, November 28–29, Hobart, Australia.

Chan, J., To, H. P. & Chan, E. (2006). Reconsidering social cohesion: Developing a definition and analytical framework for empirical research. *Social Indicators Research*, *75*(2), 273–302.

Chiu, R. L. H. (2004). Socio-cultural sustainability of housing: A conceptual exploration. *Housing Theory and Society*, *21*(2), 65–76.

Chiu, R. L. H. (2008). Editorial introduction: Urban housing policy issues in re-surging Asia. *Urban Policy and Research*, *26*(3), 245–247.

Doling, J. & Ronald, R. (2014). *Housing in East Asia: Socioeconomic and demographic challenges*. Basingstoke: Palgrave Macmillan.

Groves, R., Murie, A. & Watson, C. (2007). *Housing and the New Welfare State: Perspectives from East Asia and Europe*. Ashgate: Aldershot and Burlington.

Grzeskowiak, S., Sirgy, M. J., Lee, D. J. & Claiborne, C. B. (2006). Housing well-being: Developing and validating a measure. *Social Indicators Research*, *79*, 503–541.

Hulchansky, D. (1995). The concept of housing affordability: Six contemporary uses of the housing expenditure-to-income ratio. *Housing Studies*, *10*(4), 471–491.

Midgley, J. (2014). *Social development: Theory and practice*. 55 City Road: Sage Publications, pp. 9–13.

Oxoby, R. (2009). Understanding social inclusion, social cohesion, and social capital. *Journal of Social Economics*, *36*(12), 1133–1152.

Pawar, M. S. & Cox, D. R. (Eds.) (2010). *Social development: Critical themes and perspectives*. New York: Routledge.

Seo, B. & Chiu, R. L. H. (2014). Social cohesiveness of disadvantaged communities in urban South Korea: The impact of the physical environment. *Housing Studies*, *29*(3), 407–437.

Unger, D. G. & Wandersman, A. (1985). The importance of neighbors: The social, cognitive and affective components of neighboring. *American Journal of Community Psychology*, *13*(2), 139–169.

2 Housing policy and urban development in China

The public housing perspective

Jie Chen

Introduction

Over the past three decades, China has experienced rapid urbanization, and the urban sector has dominated the national economy. While it is true that the rapid urban growth arises mainly from the productive forces unleashed by China's economic reforms (OECD, 2012), how the institutional changes of housing regimes affect the mode of urbanization has not received sufficient attention in the literature. This paper puts forward an examination of the interactions between housing regime and urban development in China, with special attention given to the roles of public housing provision in China's urban development in catering for the needs of the low-income groups.

In the first place, the shifting of the responsibility for providing responsibility for urban housing from work units to the market is a precondition of the market-oriented reform of state-owned enterprises in the early 1980s (Shaw, 1997; Wang, Wang & Bramley, 2005; Wu, 1996), which made the revitalization of the Chinese urban economy possible. Second, the transformation of urban housing from welfare goods to commodity goods greatly helps to drive the consumption boom and economic prosperity in urban China (Chen, Guo & Zhu, 2011; Song, Chu & Chen, 2004). Further, an increasingly flexible housing market with responsive supply elasticity is a key force in propelling the rural-urban population mobility as well as inter-city labour mobility in China. In addition, the local states' revenue from the sale of residential land has played a crucial role in financing Chinese cities' infrastructure investment since the 1998 reform (Wang, Zhang, Zhang & Zhao, 2011; Ye & Wu, 2014). As this chapter will show, the great housing market boom in China is not only a consequence but also a root of China's great leap forward of urbanization in the new century. In short, the changing functions of state intervention in the Chinese housing sector have yielded far-reaching impacts on the urbanization in China. Nonetheless, the Chinese post-reform urban housing regime has put too much priority on the economic functions while largely overlooking its social functions, especially the needs of rural-to-urban migrants. This imbalance has produced severe obstacles for the sustainability of urban development in China.

The remaining sections of this paper are organized as follows: it initially explores the interaction between housing regime and urbanization in general, then investigates how housing market development positively contributes to urban development in China, and discusses the associated problems of housing market development, subsequently it explains the re-emergence of state provision of public housing to cater for the low-income groups, and finally concludes that China's recent trend of shifting the housing regime from "productivism" to "developmentalism" is to ensure the successes of the new urbanization strategy.

Housing and urban development: an analytic framework

This section briefly introduces a simple analytic framework for ways the dynamics of the housing regime may affect urban development and then provides a brief literature review of this topic.

Housing and urban development: the channels

The housing regime, the structure for housing provision, can affect urban development through several channels. Below we discuss four major channels that link the housing regime to urban development.

First, the housing regime affects the scale and institutional structure of housing construction, which provides the most fundamental physical foundation of urban expansion (Mayer & Somerville, 2000). Second, the housing regime decides how housing is produced and delivered to the laborers and their families, which directly determines the size and spatial allocation of an urban population. For example, the responsiveness of housing supply plays a crucial role in determining housing prices and in turn affects urban employment growth because the supply capacity and prices of urban housing affect the scale of cities (Glaeser, Gyourko & Saks, 2006). On one hand, there is evidence that, in areas with an inelastic housing supply, increase in labour demand would generate stronger responses in housing prices, forcing workers to ask for more wage compensation and then resulted in greater increases in labour costs, retarding employment growth (Saks, 2008). On the other hand, the elasticity of housing supply can be severely land-constrained by the geography of a city, and housing land supply also contributes to reshaping the spatial distribution of the urban population and urban economy (Saiz, 2010). Third, the funding of housing investment is highly integrated with the financing of general urban infrastructure and urban public service, which underpin the formation of urban amenity and eventually affect the relative attractiveness of cities (Gauvin, Vignes & Nadal, 2013). Fourth, the housing regime has direct impacts on the price movements of housing assets. Through households' consumption or saving behaviour (wealth effects), borrowing or lending behaviour, and asset-building or asset-consuming behaviour (reverse mortgage), the fluctuation of housing prices deeply affects the trends and business cycles of an urban economy (Iacoviello, 2005).

Housing and urban development in China: the literature review

Many newly-emerged economic bodies have experienced a continued economic boom over the last three decades, together with fast industrialization, rapid urbanization and speedy modernization. As Jim Kemeny says, housing is a fundamental dimension of the social structure (Kemeny, 1992). Investigations of the housing sectors would provide important insights into the social-economic restructurings in these countries. In this respect, housing development has been both a subject and a catalyst of urbanization.

National housing systems in East Asia share an important common goal of using housing development as a means to drive urban development and economic growth (Doling & Ronald, 2014). Through promoting asset-based welfare among home owners, East Asian governments expect that family assets can function as social security and the building up of "asset-based security" can preserve self-sufficiency and reduce the citizens' demand to develop onerous and costly welfare states. As noted by some scholars:

> home ownership was promoted as a means of, on the one hand, contributing directly to economic growth through the motor of the construction industry, and, on the other, supporting a low taxation, low public expenditure economy with minimal social protection measures through the support of the family.
>
> (Doling & Ronald, 2014)

The asset-based welfare philosophy reflects the ideology of productivist welfare capitalism. Holliday (2000) introduces the term "productivist welfare regime" by stressing that social welfare policy is typically an extension of economic policy, both subordinated to and defined by the growth-oriented economic policy in East Asia. In such a welfare regime, social rights are minimal and extend only with productivity concerns, and the state-market-family relationship is premised on overriding growth objectives (Holliday, 2000). This theory also stresses that the social (welfare) policy should be targeted at the most productive members of the society to promote economic growth (Holliday, 2000). It is a widely held belief in the region that "economic growth can be achieved by means of selective social policies to enhance labour productivity and growth" (Lee, 2008, p. 272).

In recent years, the concepts of productivism welfare and developmentalism welfare have significantly influenced studies of welfare regimes in East Asia, and housing research is no exception (Chen, Yang & Wang, 2014; Doling & Ronald, 2014; Ronald & Doling, 2010; Wang & Murie, 2011). In terms of the shared common interest in the close relationship between economic development and social policy, productivism and developmentalism appear to be synonymous. However, they stand in stark contrast to the relationship between economic policy and social (welfare) policy. While the assumption of the subordination of welfare policy to economic policy is a central key in productivism, developmentalism emphasizes "a need to integrate economic and social policy because social expenditures in the form of social investment do not detract from but contribute positively to economic development" (Midgley & Tang, 2001, p. 246). The developmentalist theory of welfare capitalism thus aims mainly to challenge the neo-liberal claim that social welfare expenditure harms the economy (Midgley & Tang, 2001).

However, complete state withdrawal from housing provision in East Asia has triggered persistent housing market turbulence, economic volatility, social fragmentation and wealth polarization in the urbanization process in this region (Ronald & Chiu, 2010). Accompanied by the increasing demand for social rights and democracy, the rapid ageing of the population, falling fertility and the erosion of the traditional family model during the rapid urbanization process, the East Asian housing model is being reshaped, and this is especially true in China.

The roles of housing market developments in China's urban development

China's urbanization process is unique in several respects. First, its scale is unprecedented in human history (United Nations, 2012): more than 300 million residents have moved from rural areas to urban areas since the turn of the century. Second, its speed is drastic in that the urbanization rate has increased by one percent point annually over the last decade. Third, the urbanization rate in China is still significantly below the average level in economies at similar development stages (Henderson, 2009). China's urbanization is also consistently lagging behind that of the industrialization rate, measured by either the ratio of non-agriculture outputs in GDP or the ratio of non-agriculture labour force in the total labour force. Fourth, there are sharp insider-outsider divides under the so-called "dual-track" urbanization in which new urban households with and without the urban *hukou* (household registration) system are experiencing much different access to local public services. Finally, it is occurring within a Communist one-party system.

Against this spectacular background, interpreting the transformation of the Chinese urban housing regime is especially challenging. China is a hybrid developing, transitional and East Asian country, yet conceptual frameworks of housing regimes and urbanization modes are still derived from predominantly Western paradigms. Frameworks adapted from other East Asian countries seem to offer the greater salience but cannot be adopted uncritically. The dynamic nature of the transformation is a further complicating factor. However, the recent radical shift in the orientation of Chinese housing policy since 2007, including the adoption of a mass public housing programme between 2011 and 2015, urgently necessitates timely reflection and new re-interpretations. The new public housing programme in post-reform urban China initially originated as part of the emergency economic stimulus package in 2008 but now forms part of the emerging new urbanisation strategy, associated with the new leadership.

The new urbanisation strategy is a component of the Chinese government's series of attempts seeking to rebalance the economy. In turn, with assistance of the new public housing programme, the China's new urbanisation strategy is prompting changes in the development of (non-housing) welfare provision, and possibly radical reforms of the *hukou* system.

The existing literature has pointed out that the launch of the reform of Chinese urban housing regime in the 1980–1990s was mainly driven by economic reasons (Man, 2011; Wang et al., 2005). The direct goal was to reduce the soaring fiscal burden on urban housing subsidies (World Bank, 1992). However, the Chinese government has for decades used the development of a commercialized housing sector as a motor to drive the urban economy (Wang & Murie, 1999).

Housing investment as growth motor of urban economy in China

By any standard, the housing sector has become a leading industrial pillar in the Chinese urban economy since the market-oriented economic reform. Since the late 1990s, real estate investment has consistently accounted for roughly 10 per cent of the Chinese GDP. Several empirical works confirm that housing investment has played a pivotal role in the Chinese economy (Chen et al., 2011; Economist Intelligence Unit, 2011). In addition, they also demonstrate that housing investment has an active role in driving China's economic fluctuations (Chen et al., 2011).

The national statistics show that the real estate industry alone created 3.7 million jobs in 2013, which was 2.5 million higher than in 2003 (NBSC, 2014). The construction industry is perhaps the industry that benefited most from the housing market boom. Most jobs in the Chinese construction industry are taken up by rural-to-urban migrants. According to a recent survey published by the National Bureau of Statistics of China (NBSC, 2014), 22 per cent of rural migrant workers were employed in the construction industry, and this number amounted to 59.7 million in 2014.

Property-led urban redevelopment

The marketization of housing provision leads to great transformations of the urban regeneration and urban redevelopment in Chinese cities. Since the early 1990s, the state-led mode of urban redevelopment has been largely replaced by a privately funded and property-led redevelopment mode (Dowall, 1994; Shin, 2009). Subordinate to the local development strategy, urban renewal has been extensively used by municipal governments as a "growth machine" (Wu, Xu & Yeh, 2006). Property-led urban regeneration reflects the emergence of the entrepreneurial and profit-seeking behaviour of local governments in Chinese cities (Yang & Chang, 2007).

Driven by diverse motivations of different levels of the entrepreneurial state as well as profit-seeking motivations of investors, pro-growth coalitions between local government, developers and government enterprise are formed and exert powerful influences on urban redevelopment (He & Wu, 2005). During the so-called "property-led regeneration", old dilapidated houses in downtown areas are expropriated, original households are relocated to suburban areas, and residential communities in downtown areas are converted to shopping centres or other more profitable projects. By doing so, local governments obtain substantial revenues and funding sources to invest in the development of urban infrastructure and thus increase the city's competitiveness.

Meanwhile, China's market-oriented reform of housing provision has also actively contributed to transform the spatial landscape of Chinese cities (Li & Wu, 2006, 2008; Wang & Murie, 2000). In particular, job-housing relationships, transport modes and commuting behaviours have been greatly changed when more and more residents access housing from the market rather than their work units (Wang & Chai, 2009).

Housing capitalization as a financial engine of urban development

Encouraging private ownership of housing is another salient component of China's great overall transition from a socialist regime to a market economy (World Bank, 1992). The emergence of a liberalized housing market and the related liberalization of land use rights were greatly instrumental in promoting a market-driven urban economy and enabling the "growth-first" urban governance. In a fast-growing economy, the commodification and capitalization of urban housing makes the housing sector easily attract speculative funding, and housing inevitably becomes a very attractive investment vehicle. At the end of 2015, about 22 per cent of banking loans in China were issued to the real estate sector, where 14 per cent went to home mortgages and roughly 8 per cent were loans of property and land development (People's Bank of China, 2015). The ratio of the outstanding balance of home mortgages as a percentage of GDP rose sharply from 1.4 per cent in 1999 to 20.6 per cent by the end of 2015 (People's Bank of China, 2015).

Further, under the unique urban land system in China, where the local states hold a full monopoly of urban land, the capitalization (financialization) of urban land leads to the capitalization (financialization) of urban land. To be exact, with a booming housing market, the local states can both receive an enormous amount of direct funding through land sale revenue and use high-valued land as collateral to borrow loans from banks (Tao, Su, Liu & Cao, 2010). The official statistics report the land sale revenue in China has grown at a spectacular speed since the early 21th century and currently is around 3–4 trillion RMB,[1] roughly accounting for one-third of local states' total incomes (National Audit Office, 2013). Additionally, through local government-backed investment units, the debt that the Chinese local states are responsible to repay reached the substantial amount of 1 trillion RMB in June 2013, and about 37 per cent of such debt is pledged on future expected land sale revenue (National Audit Office, 2013).

Supported by the funding from these sources, the entrepreneurial local states fiercely compete for profitable industrial investment at the global scale by either providing subsidies and tax incentives (Lichtenberg & Ding, 2009), or by enhancing urban amenities through large investments in local urban infrastructure (Wang et al., 2011). Thus, land-based financing of urban development has been identified as the most significant feature of China's urbanization (Ye & Wu, 2014).

The problems of housing market development in the urbanization process in China

The marketization of housing provision has greatly paved the way for the rural-to-urban migration. It has also created millions of job opportunities for migrants. Nonetheless, the rapid upsurge of housing price has also made decent housing accommodation in the cities increasing unaffordable for migrants (Chen, Hao & Stephens, 2010; Yang & Shen, 2008). Over time, the lack of affordable housing provision has posed increasingly severe challenges for the sustainability of the urbanization in China.

Despite the rapid urbanization, migration in China is still subject to many institutional restraints. The *hukou* system, the urban registration system that was introduced in 1958 (although it had its precedents), has been promulgated as an official tool to control the free movement of people between urban and rural areas (Chan & Buckingham, 2008). Administrative regulations issued in 1982 and known as "custody and repatriation" (C&R) authorized the police to detain migrants if they did not have a local residence permit (*hukou*) or temporary living permit in the city and repatriate these "illegal migrants" to the settlements where their permanent *hukou* were issued (Chan & Buckingham, 2008). Although the C&R law was formally abolished in 2003, the *hukou* system remains in force now and continues to constrain population mobility. Under the *hukou* system, migrants are largely excluded from the welfare package that is reserved for local residents, including unemployment insurance, health care, pensions, housing benefits and even their children's rights to enter local schools (PFPC, 2012).

The *hukou* system has been constantly noted as a main constraint for the housing consumption of migrants (Sato, 2006; Wu, 2004). At the national level, an official survey indicates that the home ownership ratio among migrants was just around 10 per cent in 2010 (PFPC, 2012), which is in sharp contrast with 89 per cent among permanent urban residents (NBSC, 2011). The exclusion of migrants from the local housing welfare system further exacerbates migrants' difficulties in the Chinese urban housing market (Sato, 2006). Several micro studies show that migrants in major cities are not well integrated spatially with local residents (Li & Wu, 2008). They generally live outside of the expensive downtown and cluster heavily in the cheaper suburban areas with plenty of low-skilled job opportunities and informal housing accommodation solutions. The rise of *hukou*-based residential segregation is not only a self-sorting market equilibrium but also a result of the discriminating institutional forces, as the migrants are constrained in their housing choices under the *hukou* system. For example, since 2010, in many big cities migrants without local *hukou* could not purchase local commodity housing. It is a strong signal that the Chinese urbanization mode is neither a pro-poor one nor an inclusive one.

Rural migrants are housed mainly in two broad types of housing. Many are housed in dormitories and other forms of temporary accommodation (e.g. semi-completed buildings on construction sites) provided by employers (PFPC, 2012). Others are housed in so-called 'urban village' areas that represent two of the underlying dynamics of urbanisation: the rapid expansion of cities into former farmland, and the urgent need for migrants to find affordable shelter (Wang, Wang & Wu, 2009). Former farmers owning the land of urban villages construct low-quality but often multi-story properties and rent them to impoverished migrants at low prices. Urban villages in China share many features with the shanty towns found in other developing countries (Wang et al., 2009). Most housing in urban Chinese villages is sub-standard, intensely developed, densely populated, with poor infrastructure and often illegal, in the sense that they are not recognized by any form of urban planning (PFPC, 2012; Zheng, Long, Fan & Gu, 2009;). Urban villages are often blamed by the local governments as the breeding grounds for social problems such as crime and an underground economy. Some researchers consider urban villages to be a form of slum with Chinese characteristics (Hao, Geertman, Hooimeijer & Sliuzas, 2012; Liu, He, Wu & Webster, 2010).

Public housing and urban development in China

The rationales of new public housing programs in China

It has been argued that many aspects of urban dynamics are strongly affected by local housing policy (Glaeser et al., 2006). Public housing is perhaps the most controversial housing policy as it is in direct competition with market housing and also costly to implement and maintain (Green & Malpezzi, 2003). Nonetheless, direct state provision of housing remains an important element of housing regime in many countries. As suggested by Green & Malpezzi (2003), "political support is generally stronger for programs tied more closely to the consumption of specific goods (housing, food, and medical care) than for income support".

Being a developmental state, the Chinese state envisages economic growth as the most important means to earn the political legitimacy to govern. From 1998 to 2011, China witnessed an unprecedented construction boom with more than 9,300 million square metres added to the residential stock, which made it possible to shelter the 280 million population that migrated from rural areas to urban areas. However, as Wu (2001) pointed out, China's post-reform housing policies are embedded with two interrelated but contradictory objectives: on one hand, to increase housing supply sufficiently to accommodate rapid urbanization through commodification and marketization of housing; on the other hand, to stimulate local growth through enhancing the attractions of real estate investment.

In recent years, the embedded contradiction within housing policy aggravated a number of threats to the state's political legitimacy, which include general worsening of housing affordability and rampant property speculation as well as increased macroeconomic instability. To confront these threats, since 2009 the Chinese central government has been mandating all municipalities to construct large-scale public housing projects (Wang & Murie, 2011). In spring 2011, the State Council promised to deliver 36 million units of public housing during the period of "the Twelfth Five-Year Plan (FYP)" (2011–2015). Should this objective be achieved, the proportion of the country's formal housing stock through public provision would almost double from 11 per cent to 20 per cent by the end of 2015 (MOHURD, 2011).

The structure and purposes of new public housing programs in China

On July 2015, the Ministry of Housing and Urban-Rural Development (MOHURD) announced on its official website that 32.3 million units of public housing had been started between 2011 and 2014, and another 7.4 million units were expected to be built in 2015 (Table 2.1). These 40 million new constructed public housing units, if completed, would add roughly 20 per cent to the housing stock in the urban areas of China.

The new Chinese public housing program includes a mix of 'products' for different segments of the market. Approximately 60 per cent of public housing provisions are forms of subsidized owner-occupied housing and 40 per cent are for rent (cf. Table 2.2). Nonetheless, there is no data on how much public rental housing is reserved for low-income households.

Table 2.1 The construction of public housing in China, 2011–2015

Index	*Newly Started Public Housing (within a year)*	*Completion rate of Annual Plan*	*Finished Public Housing (within a year)*	*Investment in Public Housing (within a year)*	*New Rental Subsidies*
Until	*Units*	*%*	*Units*	*Billion RMB*	*1,000 Households*
2011–09	9,860,000	98.00	–	–	215
2012–10	7,220,000	103.00	5,050,000	1,080	–
2013–11	6,660,000	100.00	5,440,000	1,120	–
2014–09	7,200,000	103.00	4,700,000	1,070	–
2015–08	5,980,000	80.00	5,060,000	920	–
2015–09	6,850,000	92.00	6,250,000	–	–

Source: The section on public housing progress on the official website of MOHURD (Ministry of Housing and Urban-Rural Development).

Table 2.2 The supply structure of Chinese public housing, 2011–2015

Type of housing	*Function*	*Proportion of program (%)*[1]
Cheap Rental Housing (LRH)	Government-owned housing leased with low rents to poorest local urban residents. Limited to 50 m^2. Financed mainly by local government, but with increasing central government component.	ca. 10%
Public Rental Housing	Government-owned housing leased with quasi-market rent to middle households, accessible for migrants. The size is usually limited to 60 m^2.	ca. 20%
Economic & Comfortable Housing (ECH)	For-sale housing targeted at low-middle income households with restricted ownership. Limited to 60 m^2 and sold at prices 30–40% below market levels. Developers' profit margins usually capped at 3–4%.	ca. 20%
Capped-Price Housing (CPH)	For-sale housing targeted at middle-income households. Limited to 90 m^2. Prices capped by local government when land sold to developer. Price is usually between ECH and market.	ca. 10%
Shanty-town Resettlement Housing (SRH)	Housing reserved to those households displaced by shanty-town clearance. The price is usually 30–40% lower than the market level.	ca. 40%

Source: The section on public housing progress on the official website of MOHURD (Ministry of Housing and Urban-Rural Development).

Note

1 These numbers are percentages of units of new housing construction.

It has been clearly stated that the primary objective of the 36-million public housing programme is to ensure the success of the new urbanization strategy (Li, 2011; MOHURD, 2011). With the slogan of "human-oriented urbanization", the new urbanisation strategy appears to represent a critical juncture in Chinese public policy. The primary objective of this strategy is to promote inclusive urban growth so that an increasing number of migrants can be finally settled in the cities (Koen, Herd, Wang & Chalaux, 2013; OECD, 2014). The strategy eventually aims to re-orientate the economy away from one based on exporting goods cheaply to the rest of the world, to one that boosts domestic consumption (Li, 2012; World Bank & DRC, 2013).

First, one may note that 40 per cent of these planned 36-millikon public housing are Shanty-town Resettlement Housing (SRH). This, however, should not be surprising as the state provision of Resettlement Housing has been a critical factor underlying the so-called property-led massive urban redevelopment in China since the mid-1990s (He & Wu, 2005; Shin, 2009). Assistance for shanty-town dwellers is linked to resettlement, which may take the form of low cost home ownership, low cost rental housing, or cash compensation. The local governments simply cannot afford for compensation to grow at the same rates as the soaring prices of market housing. Thus, the main function of SRH is to re-house expropriated urban and rural households at low costs and thus help to facilitate "growth-promoting" urban regeneration.

Second, the Public-rental Housing (PRH) program plays a special role in supporting urbanization. The central government does not set detailed rules regarding the objectives and qualifications of PRH. It leaves local governments with substantial freedom in making their own allocation policies. The eligibility criteria of PRH are much broader, and they vary substantially from city to city. For many cities, PRH is not a residual safety net for low-income households but instead a tool to support local economic development. According to guidelines on accelerating the development of PHR released by MOHURD and another six departments in July 2010 (MOHURD, 2011), PRH is introduced with the main purpose of solving temporary and interim accommodation needs of migrants, new workers and house-poor households (Li, 2011). However, local governments would more likely allocate PRH as subsidized benefits to specific target groups that they prefer, for example, local civil servants or high-educated skilled labours they want to attract and retain (Wang & Murie, 2011). The provision of PRH is thus an effective selective tool used by policy makers to retain skilled or semi-skilled workers in large cities. For example, the recent development of PRH projects in Shanghai is mainly a result of a deliberate urban development policy in line with other strategies such as city marketing and gentrification. Because there is very limited security of tenure in the private rental housing market in China (Man, 2011), PRH in the form of "gated communities" has strong attractions for the middle class (Wu, 2005), who highly value residential stability. In this respect, PRH provides an alternative to home ownership with an affordable and guaranteed leasing contract offering decent housing to the newly-emerged middle class.

Public housing and urban development: the case of Shanghai

This section illustrates the roles of public housing in China's urban development by using the Shanghai experience as a case study. It traces the recent experience of public housing development in Shanghai and focuses on the roles of PRH programme in Shanghai's recent socio-spatial dynamics. It is shown that the public housing programme in Shanghai is mainly a result of deliberate urban development policy in line with other strategies such as city marketing and gentrification. Thus, we suggest that the Shanghai municipality government appropriates the new public housing regime as an institution to buttress local economic competitiveness. The analysis is strengthened with data from a survey of PRH tenants in Shanghai.

Key concepts and major backgrounds

As argued in some of the literature, Shanghai has embraced a state-led development approach but functions as an entrepreneurial city when paving the way to reclaim its global status (He & Wu, 2005; Wu, 2003; Zheng, 2010). City marketing is one of the main features of an entrepreneurial city, and particularly for cities that have embraced global competition. Sager (2011) emphasizes the neoliberal rationale behind it: 'city marketing, promotion and branding are means for achieving competitive advantage in order to increase inward investment and tourism' and two groups of 'placer customers', together with the visitors, are usually the targets of city marketing drives: specifically:

1 inhabitants who want an attractive place to live, work and relax and
2 companies looking for a place to locate their offices and production facilities, do business and recruit employees (Sager, 2011, p. 157).

As summarized by Sager (2011, p. 157), "the creative class needs places for consumption, recreation, and living. ... Furthermore, housing the creative class requires a shift from working class quarters to hip, varied and good quality residential areas."

Important means used in city marketing include flagship projects and mega events (e.g. a Formula One race event and the well-known World Expo 2010 were held in Shanghai). In Shanghai, the city has employed various preferential policies to create an attractive image as an ideal place for industrial development and financial investment. Creative industry clusters have been tossed into a hub to host world-famous cultural and artistic events (Zheng, 2010). It is also suggested that Shanghai manifests a complicated relationship between gentrification, globalisation, and emerging neo-liberal urbanism, and the local state has played a leading role in the large scale gentrification in Shanghai, mainly through the strategic plan (He, 2010).

Public housing framework in Shanghai

In 2010, nearly twenty years after the termination of welfare rental housing, Shanghai Municipality adopted the PRH Programme and branded it as one element of the 'four in one' comprehensive public housing policies. The central idea of the 'four in one' model is to provide different types of affordable housing for different groups: the Cheap Rent Housing for the poorest households; the EAH (Shared-ownership) for the low-middle income households; the resettlement housing for the households displaced by the government; and the PRH for those who cannot afford home ownership but are also excluded from other three affordable housing programme. The PRH program is the only scheme open to non-*hukou* holders; however, they still must fulfil the eligibility criterion of possessing a long-term residence permit.

According to the Shanghai Housing Development Plan for the 12th FYP (2011–2015) published on February 7, 2012, the supply plan for public housing in Shanghai between 2011 and 2015 is one million units in total: 400,000 units for EAH; 350,000 units for relocation housing; 75,000 units for CRH; and 200,000 units for PRH (Table 2.3).

Table 2.3 The supply plan of the "four-in-one" public housing in Shanghai, 2011–2015

Types	*2011–2015 (target of net increase)*	
	Units (1,000)	*Population Coverage (%)*
Cheap Rental Housing	75	1.5 (2.6)
Economical Affordable Housing	400	4.2 (7.6)
Relocation Housing	350	3.5 (6.0)
Public Rental	200	1.2 (1.9)
Total	1,000	11.8 (19.2)

Source: Shanghai Municipal Government (2012).

Note
In the last column, numbers in parentheses refer to permanent households (those with local *hukou*) only and those outside parentheses refer to the whole resident households, including floating migrants (those without local *hukou*). The population in 2015 is the author's own estimation. For Low Rental Housing and Public Rental housing, the figures include the additions from the purchase or conversion of existing housing stocks. Definitions of each type of public housing are given in the texts.

Features of public rental housing in Shanghai

In Shanghai, the PRH companies are legal independent entities, with investments shared equally between the city and the district. The PRH Company is responsible for the investment, planning, design, administration, and management of the PRH. It is financially independent, which means that for additional costs beyond the initial investment, it must finance themselves. In this respect, the PRH Company in Shanghai resembles the municipal housing company in Sweden and social housing co-operations in the Netherlands. However, it is still unclear who will finance the operating deficit of the PRH Company if the deficit occurs.

According to the government policy statement (Shanghai Municipal Government, 2010), the principles of PRH in Shanghai can be summarized as "led by the government, supplied by multiple sectors, provided at market price, and subsidised by multiple means". Specifically, the government is responsible for the policy making, planning, organising and coordinating different sectors in implementation; both public and private sectors could be the suppliers; the rental price is at market level; and the gap between the market price and affordability should be met by the subsidy shared by the municipality and employers. It needs to be noted that although no permanent register status (*hukou*) is required, the applicant has to possess a long-term residence permit and have continuously contributed to the social insurance account for at least 12 months but there are no income limit requirements. Since summer 2013, the application for residence permits has changed into points-based system, which gives higher scores for candidates who are younger, have higher education, are higher skilled, and work in sectors determined by the government as having development priorities in remote new towns. In the following sections, we borrow the term 'talented class' (*rencai* in Chinese) to refer to the semi-skilled and skilled labourers who are welcomed (or selected) by the city of Shanghai (Shanghai Municipal Government, 2010).

In short, the new PRH programme in Shanghai is tailored for:

1 The talented class who cannot afford home ownership from the market while not eligible for other affordable housing programmes, for instance, the EAH;
2 Residents who live in overcrowded housing; this implies that home owners can also apply for PRH, as long as their construction space per person is less than 15 square metres.

It should be emphasized that the rental rates of PRH in Shanghai are only slightly lower than the private rental market price. We will elaborate on the implications of this point in later sections.

The demand for a talented class and PRH in Shanghai

Shanghai has been given a role by the state as the "dragonhead" to lead the "opening up" policy in the post-reform era, with an ambitious aim of becoming a global financial centre. A strong relation has been found between economic globalization and the marketization of the socialist system in China (Witt & Redding, 2014). Meanwhile, it has been suggested that globalization and competitive strategies are bound together to reshape the landscape of Chinese cities (Xu & Yeh, 2009). There is also a significant literature on the diverse city marketing and place promotion methods used in Shanghai (Wei, Leung & Luo, 2006; Wu, 2001; Yang & Chang, 2007).

To meet the growing demand for an entrepreneurial city, Shanghai needs more human inputs as the engine of growth. The decentralization of the Chinese fiscal regime since the mid-1990s has permitted Shanghai to embark on its entrpreneurial jouney with own fiscal resources. However, due to the rapid aging of the local population that is amplified by the strictly implmented one-child policy, the insufficient partipation of migrant contributors to the Chinese social insurance system has weakened the fiscal system. In addition, Shanghai is faced with growing competition from neighbouring cities in the Yangtze Delta for high-skilled workers.

The importance of providing decent housing to talented people has been repeatedly highlighted in government documents and meeting minutes. The 12th FYP Development Plan of Shanghai states:

> it is a crucial time for Shanghai to fulfil its goal of becoming 'four centres' and a global metropolitan area, but we are faced with many challenges … we need more innovative public policy for the talented class, and to improve the living and cultural environment for the talented class.

The slogan above is not an initiative but a formal recognition and adoption of recent practices to link housing to employment. As early as the late 1990s, many joint ventures in Shanghai bought "commodity" housing for their employees to attract capable staff (Wu, 2001). Further, another type of apartment, "the apartment for talented professionals" (*rencai gongyu*) has also emerged in Shanghai in recent years. In 2011, the Changning District collected 500 apartments units for the "talented professionals" (experts and skilled workers) mainly by adaptive reuse of vacant office buildings, hotels, and industrial buildings. The tenants can receive a heavy rent subsidy from the government (*Jiefang Daily*, 2011). However, a rent level close to market rate has excluded those low-income households from the PRH programme. From this perspective, PRH is a very selective programme with a clear target to attract the 'talented class' but gives little consideration to solving the affordability problems of those low-income migrants.

Insights from a survey of PRH projects in Shanghai

This section provides some observations from a survey of tenants living in the first two PRH projects in Shanghai (Shangjing Garden and Xinning Apartments). The two projects provide 5,100 housing units in total and have been available for rental application since March 2012. Unlike most commodity housing in China, which is unfurnished at the delivery stage, the PRH apartments were well furnished, and applicants can move in immediately after getting approved. The initial lease period is up to six years, and it is possible to renew. The PRH housing are usually two-bed apartments of around 60–70 square metres. Standard public facilities are installed in the neighbourhoods of PRH projects.

To understand who had been attracted to the new PRH projects, we provide information on the characteristics and housing satisfaction of PRH tenants, based on a survey of the residents of the two PRH in Shanghai. This survey was conducted during June-October of 2012 by the Centre for Housing Policy Studies (CHPS) at Fudan University. Roughly 2,400 tenants lived in the two projects at that moment and a randomly-chosen sample was surveyed by paper questionnaires. The final sample size is 333 in total, 128 from Shangjin Garden and 205 from Xinning Apartments.

This survey shows that most PRH tenants are middle-class: 64 per cent of survey respondents report their personal annual disposable income as higher than 60,000 RMB, 30 per cent higher than 90,000 RMB and 13 per cent higher than 120,000 RMB (note that the mean level of annual disposable income of Shanghai residents in 2011 was 36,230 RMB). Further, the PRH tenants are mostly young and middle-aged: 65 per cent of respondents are below 35 and 44 per cent younger than 30; only 14 per cent are older than 50. In addition, a high education level is one of the main features of PRH tenants: 65 per cent of respondents have received a Bachelor's degree or higher.

The survey also shows that the majority of PRH felt satisfied with the overall quality of the PRH project: 59 per cent of respondents thought PRH met their expectations and 17 per cent thought PRH exceeded their expectations;

however, 24 per cent felt PRH failed to meet their expectations. The aspects of PRH that respondents were most satisfied with include tenure security (30 per cent), housing quality (18 per cent) and community security (17 per cent). The aspects that tenants felt least satisfied with include rent rate (3 per cent), convenience to the work place (4 per cent) and layout and design (6 per cent).

Because tenure security has little protection in the private rental housing market in China (Man, 2011), PRH has strong attractions for the middle class, who highly value residential stability. Further, the high ratio of housing satisfaction among PRH tenants can also be attributed to the fact that the PRH projects are "gated communities". Wu (2005) suggests that the primary reason for the new emergence of gated communities is more about the protection of lifestyle and the identity of the middle class. It also occurs in the context 'wherein the local government fails to provide differentiated services to those who are better-off in the market transition' (Wu, 2005). In this respect, PRH provides an alternative to home ownership with an affordable and guaranteed leasing contract offering decent housing to the newly-emerged middle class.

Conclusion

China's urbanization process has entered a critical stage. While the urbanization rate in China exceeded 55 per cent in 2015, hundreds of millions of Chinese still lived in the rural region. The United Nations estimate that China will add another 200 million urban dwellers in the next three decades (United Nations, 2014). The provision of affordable housing has become a political task in China to alleviate the level of inequality and income disparity generated by market and growth-led development in the post-reform era.

The massive construction plan of 36 million units of public housing between 2011 and 2015 is widely believed to serve as an economic vehicle to counteract the shocks of the global economic downturn and a regulation tool to cool down the overheating residential property markets (Naughton, 2010; Ulrich, 2011). It is also closely connected to the adoption of the "harmonious society" development ideology in 2006 (Chen et al., 2014). Nonetheless, the main purpose of the new public housing program is to ensure the success of the new urbanisation strategy that formally launched in 2013, which is essentially a reorientation of the development strategy of the Chinese economy from export-driven to (domestic) demand-driven (World Bank & DRC, 2013). Public housing policy, as a primary urban policy, is expected to achieve social-economic equality by providing decent homes for all (Li, 2011). This paper elaborates the multiple purposes behind the public housing programmes in China. A close examination of the case of PRH in Shanghai proves that the recent revival of public housing in Chinese cities is mostly driven by economic growth motives.

In the case of Shanghai, we find the PRH programme is one measure of city marketing in order to attract new talent and involves the development of gated communities for the middle classes. In particular, the existing PRH projects help alleviate the pressure of home ownership for the "young white-collar workers" by providing a decent place to live at a price they can afford. These PRH projects resemble a temporary substitute for the home ownership in gated communities that the middle class long for. According to our survey, the PRH residents are mainly the young middle class with a high education level, a group which highly values the amenities and the privatized landscape of the gated neighbourhoods with a high level of security. However, the rents of PRH are beyond the affordability range of low-income households. With the rent level close to the market price and other conditions, PRH is a very selective programme with a clear target to attract and keep "talented professionals" to enhance the city's competitiveness. Nonetheless, more consideration should be given to low-income rural-to-urban households' housing difficulties if the housing policy's long-term aim is to provide decent housing for all.

Overall, the interpretation of the new Chinese public housing regime, or the provision structure, is challenging. In its focus, the recent public housing program serves as an engine to promote accommodating millions of low-income rural-to-urban migrants in the cities under the new urbanization strategy. Interestingly, the literature has also suggested that housing policy practice in China at the moment is actually not very different from that in Western countries at a similar stage of urbanization. For example, in many Western countries, public housing was also once developed at a similar stage of rapid industrialization to accommodate industrial workers (Malpass & Murie, 1999).

The initial post-reform public housing programme in urban China had some clear indication of productivism (Chen et al., 2014). It is still widely believed that in China "(the) state housing provision is seen as important economic a driver rather than socially necessary" (Wang & Murie, 2011). Nonetheless, according to the new doctrine of a "harmonious society" proposed by Chinese President Hu Jintao in 2006, social welfare policy should be more integrated with economic policy and therefore also become a new benchmark for ranking the success of local Chinese leaders. For example, a recent joint report released by the World Bank and the Development Research Center of the State Council of China (DRCSC) suggests that China's future version should be either the "active welfare society" or "developmental welfare" model where the society should help meet the basic needs of all (World Bank, 2012, p. 298). In this chapter, we suggest some preliminary evidence that the Chinese housing regime is shifting from "productivism" to "developmentalism". Nonetheless, as both the housing regime and urbanization process are still drastically transforming, our analysis is still at a very early stage. Ultimately, it is on its distributional consequences that a welfare or housing regimes will be judged, and in urban China uncertainty prevails.

Acknowledgement

The research is supported by the NSFC-ESRC Joint Funding (NSF7161101095) and National Science Foundation of China [NSF71173045, NSF71573166].

Note

1 1 CNY = 0.158 USD (as of November 2017).

References

Chan, K. W. & Buckingham, W. (2008). Is China abolishing the Hukou system? *The China Quarterly*, *195*(6), 582–606.

Chen, J., Guo, F. & Zhu, A. (2011). The housing-led growth hypothesis revisited: Evidence from the Chinese provincial panel data. *Urban Studies*, *48*(10), 2049–2067.

Chen, J., Hao, Q. & Stephens, M. (2010). Assessing housing affordability in post-reform China: A case study of Shanghai. *Housing Studies*, *25*(6), 877–901.

Chen, J., Yang, Z. & Wang, Y. P. (2014). The new Chinese model of public housing: a step forward or backward? *Housing Studies*, *29*(4), pp. 534–550.

Doling, J. & Ronald, R. (Eds.) (2014). *Housing East Asia: Socioeconomic and demographic challenges*. Hampshire: Palgrave Macmillan.

Dowall, D. E. (1994). Urban residential redevelopment in the People's Republic of China. *Urban Studies*, *31*(9), 1497–1516.

Economic Intelligence Unit. (2011). *Building Rome in a day: The sustainability of China's housing boom*. London: Economic Intelligence Unit.

Gauvin, L., Vignes, A. & Nadal, J. P. (2013). Modeling urban housing market dynamics: Can the socio-spatial segregation preserve some social diversity? *Journal of Economic Dynamics and Control*, *37*(7), 1300–1321.

Glaeser, E. L., Gyourko, J. & Saks, R. E. (2006). Urban growth and housing supply. *Journal of Economic Geography*, 6(1), 71–89.

Green, R. K. & Malpezzi, S. (2003). *A primer on US housing markets and housing policy*. Washington D.C.: Urban Institute Press.

Hao, P., Geertman, S., Hooimeijer, P. & Sliuzas, R. (2012). Spatial analyses of the urban village development process in Shenzhen, China. *International Journal of Urban and Regional Research*, *37*(6), 2177–2197.

He, S. (2010). New-build gentrification in central Shanghai: Demographic changes and socioeconomic implications. *Population, Space and Place*, *16*(5), 345–361.

He, S. & Wu, F. (2005). Property-led redevelopment in post-reform China: A case study of Xintiandi redevelopment project in Shanghai. *Journal of Urban Affairs*, *27*(1), 1–23.

Henderson, J. V. (2009). *Urbanization in China: Policy issues and options*. Washington, D.C.: World Bank.

Holliday, I. (2000). Productivist welfare capitalism: Social policy in East Asia. *Political Studies*, *48*(4), 706–723.

Iacoviello, M. (2005). House prices, borrowing constraints, and monetary policy in the business cycle. *American Economic Review*, *95*(3), 739–764.

Jiefang Daily. (2011, July 9). The Experiences of Solving Housing Problems for the Talented Professionals in Changning. *Jiefang Daily*.

Kemeny, J. (1992). *Housing and social theory*. London: Routledge.

Koen, V., Herd, R., Wang, X. & Chalaux, T. (2013). *Policies for inclusive urbanisation in China*. Paris: OECD.

Lee, J. (2008). Productivism, developmentalism and the shaping of urban order: Integrating public housing and social security in Singapore. *Urban Policy and Research*, *26*(3), 271–282.

Li, K. (2011). Implementing the massive-scale public housing program and improving housing policy and housing provision system step-by-step. *Qiu Shi(in Chinese)*, 53(8), 3–8.

Li, K. (2012). Deepening the strategy of expanding domestic consumption through furthering the reform and openness. *Qiu Shi(in Chinese)*, *54*(4), 3–10.

Li, Z. & Wu, F. (2006). Socio-spatial differentiation and residential inequalities in Shanghai: A case study of three neighbourhoods. *Housing Studies*, *21*(5), 695–717.

Li, Z. & Wu, F. (2008). Tenure-based residential segregation in post-reform Chinese cities: A case study of Shanghai. *Transactions of the Institute of British Geographers*, *33*(3), 404–419.

Lichtenberg, E. & Ding, C. (2009). Local officials as land developers: Urban spatial expansion in China. *Journal of Urban Economics*, 66(1), 57–64.

Liu, Y., He, S., Wu, F. & Webster, C. (2010). Urban villages under China's rapid urbanization: Unregulated assets and transitional neighbourhoods. *Habitat International*, *34*(2), 135–144.

Malpass, P. & Murie, A. (1999). *Housing policy and practice* (5th Edition). Hampshire: Palgrave Macmillan.

Man, J. Y. (Ed.) (2011). *China's housing reform and outcomes*. Cambridge, MA: Lincoln Institute of Land Policy.

Mayer, C. J. & Somerville, C. T. (2000). Residential construction: Using the urban growth model to estimate housing supply. *Journal of Urban Economics*, *48*(1), 85–109.

Midgley, J. & Tang, K. (2001). Social policy, economic growth and developmental welfare. *International Journal of Social Welfare*, *10*(2), 244–252.

Ministry of Housing and Urban-Rural Development (MOHURD). (2011). MOHURD's report of latest progress of public housing program. *Meeting of the 11th NPC (National People's Congress) Standing Committee*, October 27, 2011. Sina. Available at: http://finance.sina.com.cn/focus/rdxwbzf/index.shtml, accessed on November 28, 2011.

National Audit Office. (2013). *The audit results of national governmental debt (Audit Report No. 2013–32)*. Available at: www.audit.gov.cn/n1992130/n1992150/n1992500/3432077.html, accessed on 10 October 2017.

National Bureau of Statistics of China (NBSC). (2011). *Series reports on China's economic and social development achievements during the 11th five-year planning period (2006–2010): No. 9*. Available at: www.stats.gov.cn/ztjc/ztfx/sywcj/201103/t20110307_71321.html, accessed on 1 March 2013.

National Bureau of Statistics of China (NBSC). (2014). *The monitoring report on rural migrant workers 2013*. Available at: www.stats.gov.cn/tjsj/zxfb/201405/t20140512_551585.html, accessed on 12 May 2014.

National Population and Family Planning Commission of P.R. China (PFPC). (2012). *Report on China's migrant population development 2012*. Beijing: China Population Press.

Naughton, B. (2010). The turning point in housing. *China Leadership Monitor*, 33(1), 1–10.

Organisation for Economic Co-operation and Development (OECD). (2012). *China in focus: Lessons and challenges*. Paris: OECD.

Organisation for Economic Co-operation and Development (OECD). (2014). *China: Structural reforms for inclusive growth*. Paris: OECD.

People's Bank of China (2015). *The 2015 report on the use of RMB credit by financial agencies*. Available at: www.pbc.gov.cn/eportal/fileDir/defaultCurSite/resource/cms/2016/01/2016011818543399609.htm, accessed on 6 February 2016.

Ronald, R. & Chiu, R. L. H. (2010). Changing housing policy landscapes in Asia Pacific. *International Journal of Housing Policy*, *10*(3), 223–231.

Ronald, R. & Doling, J. (2010). Shifting East Asian approaches to home ownership and the housing welfare pillar. *International Journal of Housing Policy*, *10*(3), 233–254.

Sager, T. (2011). Neo-liberal urban planning policies: A literature survey 1990–2010. *Progress in Planning*, *76*(4), 147–199.

Saiz, A. (2010). The geographic determinants of housing supply. *Quarterly Journal of Economics*, *125*(3), 1253–1296.

Saks, R. (2008). Job creation and housing construction: Constraints on metropolitan area employment growth. *Journal of Urban Economics*, *64*, 178–195.

Sato, H. (2006). Housing inequality and housing poverty in urban China in the late 1990s. *China Economic Review*, *17*(1), 37–50.

Shanghai Municipal Government. (2010). The opinions of developing public rental housing in Shanghai. Available at: www.shjjw.gov.cn/gb/node2/n8/n78/n716/u1ai174048.html, accessed on 4 September 2010.

Shanghai Municipal Government (2012) *Shanghai housing development plan for the twelfth five-year planning period 2011–2015*. Shanghai: Shanghai Municipal Government.

Shaw, V. N. (1997). Urban housing reform in China. *Habitat International*, *21*(2), 199–212.

Shin, H. B. (2009). Residential redevelopment and the entrepreneurial local state: The implications of Beijing's shifting emphasis on urban redevelopment policies. *Urban Studies*, *46*(13), 2815–2839.

Song, S., Chu, G. S. F. & Chen, X. (2004). Housing investment and consumption in urban China. In A. Chen, G. G. Liu, & K. H. Zhang (Eds.) *Urbanization and Social Welfare in China*. London: Ashgate Publishing Limited, pp. 87–106.

Tao, R. Su, F., Liu, M. & Cao, G. (2010). Land leasing and local public finance in China's regional development: Evidence from prefecture-level cities. *Urban Studies*, *47*(10), 2217–2236.

Ulrich, J. (2011, September 26). China's Affordable Housing Program is Picking up Momentum. *CNBC*. Available at: www.cnbc.com/id/44679352, accessed on 10 October 2016.

United Nations. (2012). *World urbanization prospects: The 2011 revision*. New York: Population Division, Department of Economic and Social Affairs, United Nations.

United Nations. (2014). *World urbanization prospects: The 2014 revision*. New York: Population Division, Department of Economic and Social Affairs, United Nations.

Wang, D. & Chai, Y. (2009). The jobs-housing relationship and commuting in Beijing, China: The legacy of Danwei. *Journal of Transport Geography*, *17*(1), 30–38.

Wang, Y. P. & Murie, A. (1999). Commercial housing development in urban China. *Urban Studies*, *36*(9), 1475–1494.

Wang, Y. P. & Murie, A. (2000). Social and spatial implications of housing reform in China. *International Journal of Urban and Regional Research*, *24*(2), 397–417.

Wang, Y. P. & Murie, A. (2011). The new affordable and social housing provision system in China: Implications for comparative housing studies. *International Journal of Housing Policy*, *11*(3), 237–254.

Wang, Y. P., Wang, Y. & Bramley, G. (2005). Chinese housing reform in state-owned enterprises and its impacts on different social groups. *Urban Studies*, *42*(10), 1859–1878.

Wang, Y. P., Wang, Y. & Wu, J. (2009). Urbanization and informal development in China: Urban villages in Shenzhen. *International Journal of Urban and Regional Research*, *33*(4), 957–973.

Wang, D., Zhang, L., Zhang, Z. & Zhao, S. X. (2011). Urban infrastructure financing in reform-era China. *Urban Studies*, 48(14), 2975–2998.
Wei, Y. D., Leung, C. K. & Luo, J. (2006). Globalizing Shanghai: Foreign investment and urban restructuring. *Habitat International*, 30(2), 231–244.
Witt, M. A. & Redding, G. (2014). China: Authoritarian capitalism. In M. A. Witt & G. Redding (Eds.) *Oxford handbook of Asian business systems*. Oxford: Oxford University Press, pp. 11–32.
World Bank. (1992). *China: implementation options for urban housing reform*. Washington D.C.: The World Bank.
World Bank. (2012). *China 2030: Building a modern, harmonious, and creative high-income society*. Washington, D.C.: The World Bank.
World Bank & DRC. (2013). *China 2030: Building a modern, harmonious, and creative high-income society*. Washington, D.C.: The World Bank.
Wu, F. (1996). Changes in the structure of public housing provision in urban China. *Urban Studies*, 33(9), 1601–1627.
Wu, F. (2001). Housing provision under globalisation: a case study of Shanghai. *Environment and Planning* A, 33(10), 1741–1764.
Wu, F. (2003). The (post-) socialist entrepreneurial city as a state project: Shanghai's reglobalisation in question. *Urban Studies*, 40(9), 1673–1698.
Wu, F. (2005). Rediscovering the "gate" under market transition: From work-unit compounds to commodity housing enclaves. *Housing Studies*, 20(2), 235–254.
Wu, F., Xu, J. & Yeh, A. G. O. (2006). *Urban development in post-reform China: State, market, and space*. London: Routledge.
Wu, W. (2004). Sources of migrant housing disadvantage in urban China. *Environment and Planning* A, 36(7), 1285–1304.
Xu, J. & Yeh, A. G. O. (2009). Decoding urban land governance: State reconstruction in contemporary Chinese cities. *Urban Studies*, 46(3), 559–581.
Yang, Y. R. & Chang, C. H. (2007). An urban regeneration regime in China: A case study of urban redevelopment in Shanghai's Taipingqiao area. *Urban Studies*, 44(9), 1809–1826.
Yang, Z. & Shen, Y. (2008). The affordability of owner occupied housing in Beijing. *Journal of Housing and the Built Environment*, 23(4), 317–335.
Ye, L. & Wu, A. M. (2014). Urbanization, land development, and land financing: Evidence from Chinese cities. *Journal of Urban Affairs*, 36(S1), 1–15.
Zheng, J. (2010). The "entrepreneurial state" in "creative industry cluster" development in Shanghai. *Journal of Urban Affairs*, 32(2), 143–170.
Zheng, S., Long, F., Fan, C. C. & Gu, Y. (2009). Urban villages in China: A 2008 survey of migrant settlements in Beijing. *Eurasian Geography and Economics*, 50(4), 425–446.

3 The security-based public housing policy of Hong Kong

A social development interpretation

Rebecca L. H. Chiu, Mandy H. M. Lau and Bokyong Seo

Introduction

The evolution of the public housing policy of Hong Kong has been well discussed and documented (see, for example, Castells, Goh & Kwok, 1990; Chiu 2003), but no one has interpreted the changes from a social development perspective or explained the perennial dominance of the public rental sector despite Hong Kong's reputation as the freest economy in the world. The public rental housing programme can be traced back to the resettlement schemes in the 1950s, the low-rental housing schemes in the 1960s, and the more comprehensive Ten Year Housing Programme of the early 1970s. The subsidized Home Ownership Scheme (HOS) was introduced in the mid-1970s. Since then, these two schemes have constituted the core housing programmes of Hong Kong's housing policy, supplemented by expedient interventions in the private housing sector as market conditions warrant. The two housing schemes have lasted until today except for the suspension of HOS from 2002 to 2010. Other subsidized housing schemes have been introduced, such as subsidized home loan schemes, rental allowance schemes and housing schemes for middle-income families, but they were generally short-lived or piecemeal. Over the years, despite the effort to promote home ownership at some critical time periods, the subsidized housing policy of Hong Kong has remained security-based rather than asset-based, i.e., rental housing-led but not home ownership-led, in contrast to the other economic power in Asia, Singapore.

The economic and political factors, or in short, Hong Kong's political economy, influencing the evolution of housing policy in Hong Kong have been well expounded (Castells et al., 1990; Chiu, 2001, 2003). Subsidized rental housing started with the need to clear squatter land and to supply cheap labour for the burgeoning manufacturing activities in the 1950s and 1960s. HOS was introduced in 1976 to meet the rising demand for home ownership as wages grew in tandem with continued economic development and diversification. The rapid shift of the manufacturing industry to China since the 1980s nonetheless did not lead to reductions in the provision of public rental housing (PRH). However, the economic conditions in the late 1990s sabotaged the subsidized home ownership drive despite its political importance. Politics started to play a key role in housing policy in the late 1960s, following the outbreak of large-scale riots in 1967 triggered by the Cultural Revolution on the Mainland. To help reduce the conflicts between the colonial government and the populace, an ambitious, more structured, people-centred and comprehensive Ten Year Housing Programme focusing on rental housing was introduced in 1972. As well, in the transition to become a Special Administrative Region (SAR) after 1997, in 1993, the colonial government for the first time set a target of raising the home ownership rate among households from 43 per cent to 60 per cent by the end of its rule in 1997, for the purpose of enhancing social stability (Census and Statistics Department, 1992; Hong Kong Housing Authority, 1993). The targets were not met despite the beefing-up of the Home Purchase Loan Scheme and the Sandwich Class Housing Loan Scheme, and the introduction of the Sandwich Class Housing Main Scheme (including a loan scheme and a construction scheme), as the buoyant economy had caused housing price hikes, and development and construction projects needed more time to complete.

Drawing lessons from the Singapore experience, the first SAR government of 1997 launched an even more ambitious policy of raising the home ownership target to 70 per cent and introduced or expanded a number of new housing schemes. The reason was that home ownership is crucial for nurturing a sense of belonging and maintaining social stability (Tung, 1997). Nonetheless, the crackdown on the economy by the Asian Financial Crisis of 1998 led to early termination of all the efforts, including the suspension of the long-term HOS in 2002 and the Tenant Purchase Scheme in 2003. What remained was the public rental housing programme. The second SAR government (2002–2007) explicitly announced that helping citizens with their home purchases was not its responsibility. The third (2007–2012) and fourth (2012–2017) SAR government and the current fifth government have expressed intentions to promote home ownership but have been restrained by the under-supply of land. This restraint is commonly attributed to the market-oriented land formation and land supply policy of the second SAR government. As the sole land owner in the city and as the only supplier of new land, the government stopped the formation of new land in order not to further pressurize the plummeting property market in the early- to mid-2000s. Thus, overall, from 1954 to the present time, the public housing policy of Hong Kong has remained security-based or rental housing-led, despite the intermittent government efforts to enhance home ownership or to switch to an asset-based or home ownership-led public housing policy. In 2017, 29 per cent and 16 per cent of the population live in PRH and HOS, respectively; that is, a total of 45 per cent of the population currently live in the public housing sector.[1] Before ending the political-economic discussion, a hidden factor influencing policy changes has to be pointed out. Despite its return to China after June 1997, Hong Kong remains an autonomous region without having to follow the national housing policy directives and the five-year planning cycles. Thus, unlike cities in Mainland China, Hong Kong's housing policy is formulated and revised entirely based on local situations within the regional and international economic contexts. Cyclical changes owing to national policy or national planning are avoided, enhancing policy stability and relevance to local situations.

The social factors and the social implications of Hong Kong's rental housing-led policy have been much less well studied. Chiu (2005) examined the housing outcomes of the public housing policy from the social development perspective but did not explain the evolution of policy or its broader social implications. Forrest, La Grange & Yip (2002) investigated the sense of neighbourhood in Hong Kong's housing estates, including those in the public sector, but this is only one aspect of the social outcomes of the public housing policy. This chapter aims to fill the gap: it elucidates not only the social contexts and changes that have given rise to the dominance and the limitations of PRH but also the social wellbeing engendered by the public rental housing policy, including housing wellbeing, sense of community and post-retirement protection for the low-income families.

Following this section, the chapter discusses the social contexts and social changes that necessitate the continuation of a security-based housing policy, including demographic changes, population policy and post-retirement welfare. It then canvasses the social implications of the dominance of public rental housing estates as a residential form for lower-income families, supported by the outcomes of a questionnaire survey of 600 samples undertaken in large housing estates in the city, focusing on the social wellbeing of the residents and its relation to the urban form of Hong Kong. The last section is devoted to the housing wellbeing of the lower-income families who fall out of the subsidy net or wait to be admitted into subsidized housing, as reflections of the inadequacies of an in-kind-led housing subsidy policy in a market economy, as opposed to a cash-led policy.

A security-based housing policy: social contexts and social changes

As an autonomous city that has a negligible rural sector but a political border and a population policy with tight immigration control,[2] Hong Kong has a demographic structure and more predictable trends than many other cities that are subject to uncontrollable inter-city and rural-urban migration, such as the neighbouring city of Shenzhen. Equally, with a negligible rural sector, Hong Kong cannot rely on it to provide cheap housing to the low-income groups as the urban villages in Shenzhen do. Thus, there is a greater need for subsidized housing provision by the government. In the projection of future housing demand under the 2014 Long Term Housing Strategy, the government has focused on demographic changes and the inadequately housed households and has estimated the impacts of different future economic scenarios on new household formation. The projection of future demand for subsidized housing was based on demand and aspiration for subsidized rental and sale flats and housing market conditions (Transport and Housing Bureau, 2014). Thus, demographic changes, particularly household formation trends, including the tendency of the elderly and the young adults to live independently, are important factors to determine the supply of PRH in quantum terms and flat types to meet the needs of changing household sizes, as discussed later.

The actual supply required will still depend on the prevailing eligibility criteria and the allocation priorities. The eligibility criteria comprise a means test in terms of whether household income and assets can meet household expenditure, a 5 per cent saving rate of household income and the length of residency in Hong Kong. The first factor includes housing expenditure and non-housing expenditure, which are calculated using the outcome of government household surveys and market rental trends. Families whose incomes do not meet the projected household expenditure of respective household sizes will be eligible, and the income limit is adjusted annually. The residency eligibility requires that not less than half of the family members have stayed in Hong Kong for seven or more years. The income and asset eligibility criteria are an affordability measure to ensure two things. On one hand, subsidies go to the needy families; on the other hand, the criteria provide a scientific method to ensure the inclusion of the needy in the subsidy net. Given the income eligibility for PRH allocation for three-person households and household income distribution in 2006, 2011 and 2016 (Census and Statistics Department, 2017a; Hong Kong Housing Authority, 2006; The Government Information Office, 2011, 2016), the income eligibility consistently includes households of the lowest 30 to 40 per cent, meaning that market housing in Hong Kong is unable to fulfil the needs of 30 to 40 per cent of its population under Hong Kong's free economy policy. The residency requirement can be interpreted as a priority measure in view of the limited resources to meet social needs. Whether it is equitable and discriminatory is certainly debatable.

While the means test is by and large a scientific method of defining housing subsidy needs, social trends and social policies undermine its usefulness in controlling demands. As shown in Figure 3.1, singleton households as an applicant group have been continuously expanding in absolute and relative terms until the last two years (2016–2017); in particular, young singletons below the age of 35 account for nearly half of this group. As reflected in the census data, not only have non-elderly singleton families increased, but elderly-only households (aged 60 plus) have also increased from 855,883 in 1981 to 1,250,263 in 2011, reflecting the social trends of young adults and elderly people tending to live on their own (Census and Statistics Department, 2013). This trend on one hand increases household numbers and, on the other, reduces household income, fostering greater demand for subsidized housing. Another important cause of the increasing demand for PRH is the implementation of the Minimum Wage Policy of May 2011. As the policy led to wage rises of manual labourers such as cleaners and security guards, who are the targets of the PRH policy, the Hong Kong Housing Authority raised the income eligibility limit by lifting the savings rate of household income to 15 per cent as a one-off measure in 2011. The subsequent upward adjustment according to inflation lengthened the waiting list and waiting time, as shown in Figure 3.2. These increases are also results of the

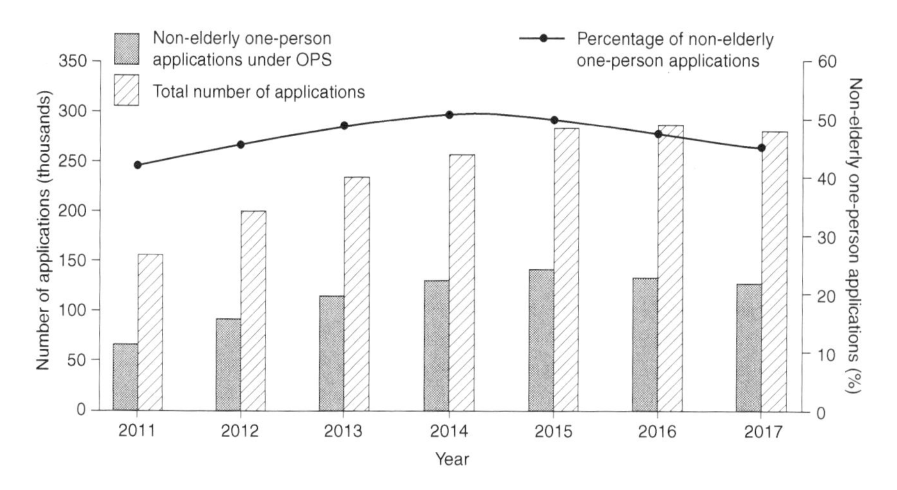

Figure 3.1 Number of applicants for public rental housing in Hong Kong (2011–2017).
Source: Hong Kong Legislative Council (various years).

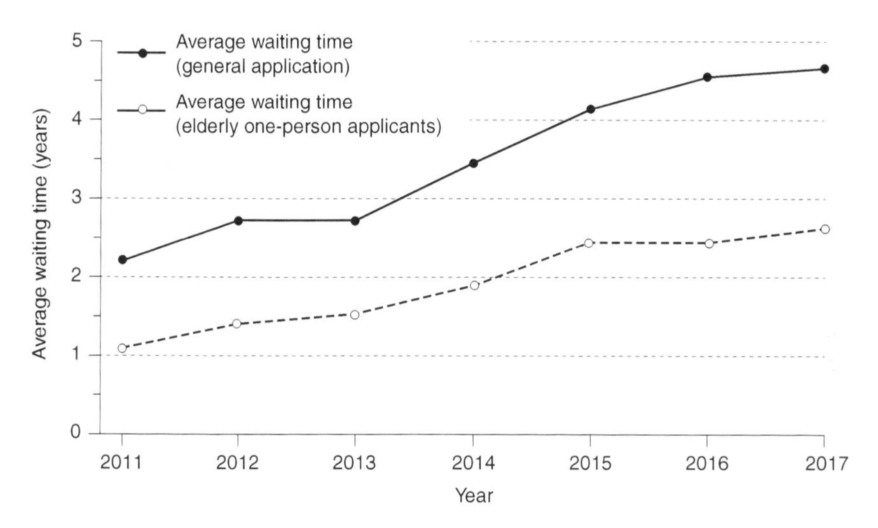

Figure 3.2 Waiting time for public rental housing allocation (2011–2017).
Source: Hong Kong Legislative Council (various years).

inability of the Hong Kong Housing Authority to increase housing supply correspondingly due to land supply restrictions as discussed.

Nonetheless, the most important factors that necessitate the continuation of a security-based housing strategy are the ageing trend and the lack of a policy to protect post-retirement welfare. Hong Kong has become an aged community since 2014, with its population cohorts above 65 years old reaching 15.5 per cent, and this cohort is expected to reach 28 per cent by 2031 and 32 per cent by 2041 because of the post-World War II Baby Boomer effect, as in many other cities. The elderly dependency ratio will decrease from 4.7 in 2014 to 2.0 in 2034 and further to 1.8 in 2041 (Chief Secretary's Administration Office, 2015). Nonetheless, Hong Kong has no comprehensive public pension system, and the contribution rate of the Mandatory Provident Fund is only 5 per cent of salary subject to a maximum contribution of HK$1,500[3] per month, thus providing very limited post-retirement protection to the elderly. There is provision of a means-tested living allowance, the Comprehensive Social Security Allowance, for the very poor, and non-means tested allowances of tokenistic amounts: Old Age Allowance, Old Age Living Allowance, and Disability Allowance. The allowances are thus either stigmatized or insignificant in providing financial support to the elderly. PRH becomes an important hedge for old age as the rent is only 25 per cent of market rent and because the Welfare Department pays the rent for recipients of Comprehensive Social Security Allowance. Unsurprisingly, elderly residents in older housing estates could reach as high as 40 per cent of the total population. As a general rule, households in public housing are subject to income checks after two years of occupancy; if found to exceed the subsidy limit, they are liable to pay additional rent or be vacated. However, elderly residents would be able to pass the check as they would normally have nil or low income.

Further, unlike private housing in the lower-end of the housing market, the design of newer public housing estates follows the principle of barrier-free accessibility, and the housing units adopt the universal design concept, both of which are more conducive to elderly living (Hong Kong Housing Authority, 2016). Residents of older housing estates built before the implementation of these design principles could request modifications to meet the contemporary design standards. Another advantage of public housing estates is its general adoption of the ageing-in-place concept, by which the estates are usually equipped with the necessary facilities and services to meet the daily needs of the elderly, including public open space and covered walkways to enhance social interaction and mobility. Also, as long-term residents, the elderly usually have built up their social communities and networks over the years. Even in redeveloped housing estates, the original community is generally preserved as the Hong Kong Housing Authority adopts the principle of re-allocation of redeveloped flats to the original residents or placing these people in estates close-by. Thus, PRH serves a very important welfare function to support the low-income elderly people, not just making up for the lack of a comprehensive pension system, but also provides a stable and socially amenable living environment. This function is increasingly important as the Hong Kong society ages rapidly. Thus, PRH serves as an important security-based welfare policy not only promising tenancy security for the low income families, but also ensuring a safer and more sociable living environment for the elderly to age in place. The social function of PRH for the low-income families is further investigated in the next section.

PRH and social wellbeing in Hong Kong

With particular emphasis on the elderly, the literature examines the correlation between housing and residents' social wellbeing in two related aspects (Costa-Font, 2013). One concerns the role of housing tenure. Many argue that home owners tend to be more satisfied with their living environment and are more involved in neighbourhood associations and community activities than renters are (Cox, 1982; Diaz-Serrano, 2009; Hu, 2013). The other pertains to the impact of the housing environment, in that it affects satisfaction, community involvement, neighbouring and neighbourhood mobility (Cattell, Dines, Gesler & Curtis, 2008; Lawton, Nahemov & Teaff, 1975; Phillips, Siu, Yeh & Cheng, 2005). In Hong Kong, public rental housing has been provided under the objectives of caring for people and creating a safe, well-maintained living environment. The PRH policy has also aimed to develop public rental housing estates in a way that is socially sustainable and environmentally friendly, and accessible by people of all ages and physical conditions. Recently, it has been increasingly emphasised that public rental housing should enable elderly residents to age in a familiar and fully manageable environment (Hong Kong Housing Authority, 2016). In this study, a face-to-face questionnaire survey was conducted in April to June 2014 to examine how these two factors interacted in Hong Kong's PRH and other housing estates. The sample comprised 600 respondents sourced from eight housing estates: two public rental housing estates (Sau Mau Ping Estate, Mei Tin Estate), one subsidized owner-occupier estate (Kwong Ming Court) and five private housing estates (Mei Foo Sun Chuen, City One, Kingswood Villas, Island Resort, Royal Ascot). Their psychological and social wellbeing was measured in terms of sense of belonging, sense of security, residential satisfaction, relationship with neighbours and participation in housing management. Housing environment was measured in terms of residents' assessment of the walkability within the estate (e.g. not too many stairs, sufficient handrails), quality of estate environment, convenience of recreational facilities/open spaces, shopping mall and public transport.

The survey results showed that the psychological and social wellbeing of PRH tenants was favourable as 77 per cent felt belonging to the community of their housing estates, and 84 per cent felt safe in the estates. Further, 78 per cent were satisfied with their living environment, and 80 per cent were happy with their relationship with their neighbours. Despite the generally low participation in housing management across the eight estates, PRH tenants in fact had relatively higher level of participation (9 per cent) than private home owners (6 per cent). The survey also found that about 86 per cent of PRH tenants expressed their satisfaction with the walkability within the estates and 82 per cent were happy with the overall quality of their estate environment. Moreover, 86 per cent and 81 per cent found the recreational facilities/open spaces and retail facilities convenient, respectively; 65 per cent found the public transportation facilities conveniently located near their estates.

In order to control for influence from potential external factors, we conducted multiple linear regressions with residents' social wellbeing as dependent variables and used respondents' socio-demographic conditions, different housing tenure types (private housing owners, private housing renters, public housing owners, public housing renters) and housing estate environment as independent variables. The regression analysis results reaffirm that there was little evidence that the psychological and social wellbeing of PRH residents is less satisfactory than that of private housing owners (Table 3.1). Specifically, the tenants' sense of belonging, residential satisfaction and relationship with neighbours were not significantly different from those of private home owners, and their participation in management was stronger than other tenure types even after controlling for other variables. Their sense of security was nonetheless lower, possibly due to the higher proportion of the unemployed among them as the average unemployment rates in PRH, subsidized owner-occupied housing estates and private housing estates were 52 per cent, 40 per cent and 34.8 per cent, respectively.

Further, the overall quality of the living environment generally and residents' perception of walkability within the estates specifically were found to be positively correlated with almost all indicators of social wellbeing, regardless of housing tenure types. This means that, first, the quality of the living environment did contribute to social wellbeing; and second, the living environment of PRH estates contributes to residents' social wellbeing in ways similar to the contribution of other housing estates to their residents' wellbeing. Moreover, the positive correlation between easy access to recreational and public transportation facilities and residents' sense of belonging, satisfaction with the living environment and relationship with neighbours means that convenience in the physical environment contributes to social and psychological wellbeing.

The favourable survey results are attributed largely to the estate design principles of achieving a healthy and green environment, ageing-in-place and a people-oriented living environment in the public housing estates (Hong Kong Housing Department, 2010; Hong Kong Institute of Architects, 2012). The estates are usually well connected to public transportation nodes such as public bus stops, subway stations and public light bus stops. They also have spacious public open spaces for social interaction and are equipped with sufficient retail and community facilities. The financial viability of providing a wide range of facilities and services also owes to the economy of scale in the high-density living environment resultant from Hong Kong's decentralized but concentrated development strategy. It thus can be concluded that the favourable housing conditions of PRH estates in Hong Kong, to some degree, offset the tenants' less favourable socioeconomic conditions, which may affect their social wellbeing. In short, the PRH policy as a security-based welfare policy not only guarantees the security of housing in terms of providing housing of a basic standard, but also a liveable environment that is conducive to low-income tenants' psychological and social wellbeing.

The case of lower-income tenants in the private sector is a different story. As indicated in Table 3.1, the sense of belonging and their relationships with

Table 3.1 Linear regression: factors of residents' social wellbeing in Hong Kong (n = 600)

	Sense of belonging	*Sense of security*	*Residential satisfaction*	*Relationship with neighbours*	*Participation in housing management*
Constant	0.972***(0.253)	1.459***(0.224)	0.674**(0.223)	0.605(0.322)	–0.312(0.328)
Gender (male = 1)	–0.009(0.050)	–0.116**(0.044)	–0.064(0.044)	–0.043(0.063)	0.012(0.060)
Length of residency	0.008*(0.003)	0.004(0.003)	0.006*(0.003)	–0.010*(0.004)	0.004(0.004)
Age (ref. 20-39)					
40–59	0.103(0.060)	0.001(0.053)	0.030(0.053)	–0.037(0.076)	0.072(0.072)
60 or above	0.079(0.088)	0.023(0.078)	–0.008(0.078)	–0.128(0.112)	0.318**(0.104)
Education (ref. primary or below)					
Secondary	0.001(0.091)	–0.071(0.081)	0.088(0.081)	–0.136(0.116)	0.107(0.109)
Tertiary or above	0.016(0.107)	–0.083(0.095)	0.070(0.094)	–0.121(0.136)	0.236(0.127)
Housing tenure (ref. private owners)					
Private renters	–0.200*(0.094)	0.058(0.083)	0.005(0.083)	–0.317**(0.119)	0.140(0.107)
Public owners	–0.093(0.094)	–0.057(0.083)	–0.038(0.083)	0.109(0.119)	0.008(0.169)
Public renters	–0.073(0.068)	–0.159**(0.060)	–0.069(0.060)	–0.016(0.087)	0.224**(0.081)
Estate walkability	0.147***(0.042)	0.131***(0.037)	0.129**(0.037)	0.140**(0.053)	–0.051(0.057)
Estate environment quality	0.297***(0.055)	0.328***(0.049)	0.336***(0.049)	0.451***(0.071)	0.225**(0.069)
Convenience of recreational facilities	0.159***(0.036)	–0.001(0.032)	0.024(0.031)	0.202***(0.045)	0.064(0.047)
Convenience of shopping mall	–0.131**(0.050)	0.081(0.045)	0.057(0.045)	0.020(0.064)	0.098(0.062)
Convenience of public transport	0.146***(0.041)	0.048(0.036)	0.223***(0.036)	0.041(0.052)	0.010(0.048)
Adjusted R^2	0.206	0.191	0.291	0.154	0.055
F	11.914***	10.915***	18.290***	8.652***	3.032***

Notes
*p<0.05, **p<0.01, ***p<0.001
Unstandardised coefficients (standard errors).

neighbours among private housing renters were lower than private housing owners as well as public housing renters. Financially, the median rent-to-income ratio of public rental housing (9.3 per cent) is considerably lower than that of private housing (30.7 per cent) (Census and Statistics Department, 2017b). The next section investigates the security and affordability issues of private rental housing, especially that of the lower-income families.

Housing security and housing affordability in the private rental sector

The public rental sector certainly enhances the housing security of those who have access to public housing. What tends to be neglected, however, is the fate of those who are left behind – those who have been queuing for public housing for many years, or those who have exceeded the income and asset limits for public housing but are unable to afford home ownership and those who do not meet the residency requirements. These households tend to rely on the private rental sector to meet their housing needs – a sector that is less secure and less affordable.

Housing security in the private rental sector can be broadly divided into two dimensions. First, security of tenure, which is what Hulse & Milligan (2014) refer to as '*de jure*' security, meaning security that is available through legal provisions. Second, rental affordability, which is what Hulse & Milligan refer to as '*de facto*' security, meaning security that is actually experienced by tenants, depending on whether they can afford rents both initially and as their tenancy progresses (p. 641). What is the level of '*de jure*' and '*de facto*' security for private renters in Hong Kong, and how has that changed over the years?

Security of tenure refers mainly to protection of tenants against 'unjust' evictions by landlords. Provision of security of tenure for private tenants in Hong Kong was previously made available under Part IV of the Landlord and Tenant (Consolidation) Ordinance. Part IV was introduced back in 1981, when insufficient supply of domestic accommodation in Hong Kong led to significant rent increases upon tenancy renewal, thereby weakening the bargaining power of tenants. It was perceived necessary to protect tenants from the risk of eviction by landlords; thus Part IV of the Ordinance restricted landlords from repossessing their properties even at the end of the tenancy period if the tenants were willing to pay the prevailing market rent. In the event that landlords wished to repossess their leased premises due to redevelopment, self-occupation or non-payment of rent, they would have to apply for permission from the Lands Tribunal.

However, security of tenure was removed around twenty years after it was introduced, when the Legislative Council approved the Landlord and Tenant Consolidation (Amendment) Bill in 2004. In other words, since 2004, the legislation in Hong Kong has not afforded private tenants protection against unjust eviction by landlords. Indeed, when we examine the debates surrounding deregulation of the private rental sector in the early 2000s, as well as the debates about possible re-regulation that emerged in the late 2000s and early 2010s, it becomes clear that successive governments have adopted a minimal-intervention approach, which has indirectly contributed to the growing problem of housing insecurity in the private rental sector.

Deregulation: removal of security of tenure

In November 2002, the Secretary for Housing, Planning and Lands announced through the Housing Policy Statement that a review of the Landlord and Tenant (Consolidation) Ordinance would be conducted, with the intention of enabling the private rental market to operate more freely. The Government's preferred option was to remove security of tenure for all tenancies, irrespective of the rateable values of the properties, or whether they were new or existing tenancies (HKSAR Government, 2003a, para. 10–12).

The proposal to remove security of tenure was not a stand-alone proposal, but an integral part of the nine measures proposed in the Housing Policy Statement to revive the stagnant property market. The Government was not particularly interested in an intellectual debate about the pros and cons of security of tenure provisions; rather, the assumption was that removal of security of tenure was naturally justified by prevailing economic conditions. These beliefs were echoed by pro-deregulation councillors in the Legislative Council, who argued that rent levels had fallen in the early 2000s, and that there was no problem with housing supply at the time (Legislative Council, 2004, p. 7537).

Anti-deregulation legislators, however, emphasized the continued importance of security of tenure in a highly volatile property market, especially considering the recurring nature of affordability and insecurity problems in the private rental sector.

> ... if security of tenure is shifted onto the shoulders of the market, actually a huge loophole will be opened. Today, the Government has not made any promise, but I wish to tell all Members and I wish to tell the Government – I call this a warning to the Government – in a couple of years, this problem is going to haunt the Government and the housing problems of Hong Kong people will not be resolved.
>
> (Legislative Council, 2004, p. 7558)

This is similar to responses made by some professional bodies to the Government's consultation paper on removal of security of tenure. These respondents questioned whether it was appropriate to remove protection for tenants based on prevailing economic conditions. For example, some respondents highlighted the cyclical nature of the local property market, and the tendency for problems in the private rental sector to recur.

> The present state of the market will not continue forever. When, or if, there is again a rental boom, the demand for security of tenure may return.
>
> (HKSAR Government, 2003b, para. 4.5)

Nevertheless, such observations did not evolve into further debates about retaining security of tenure provisions as a long-term policy tool. The Landlord and Tenant (Consolidation) (Amendment) Bill was eventually passed by the Legislative Council in June 2004, which resulted in removal of security of tenure for all domestic tenancies.

Rental increase, affordability and housing quality

Removal of security of tenure alone might not have caused serious problems, had it not been compounded by the effects of removal of rent control. If tenants were able to afford prevailing market rents, they would not be so vulnerable to displacement caused by rental affordability problems. This was not the case in Hong Kong, since rent controls had already been removed in the late 1990s.

Prior to 1998, private renters living in the older, pre-war stock and some of the post-war stock were shielded from substantial rent increases though legislative provisions on rent control. Rent control was first introduced in Hong Kong in 1921 through the Rents Ordinance 1921, due to housing shortages caused by a large influx of refugees from Mainland China (Bradbrook, 1977). After the Second World War, the Landlord and Tenant Ordinance 1947 was enacted, and it was subsequently consolidated with other ordinances to form the Landlord and Tenant (Consolidation) Ordinance 1973. The Ordinance specified maximum permissible rent increases, which were gradually raised throughout the 1970s and 1980s. In May 1992, the Government formally announced the Landlord and Tenant (Consolidation) (Amendment) Bill 1992, which proposed the phasing out of rent controls by the end of 1994. Following extensive debates in the Legislative Council, the Government deferred abolition of rent control for another two years, until 1996. Rent control was eventually abolished in 1998, following a two-year delay requested by councillors.

The removal of rent control and security of tenure was based on the assumption of falling rents in the private sector and anticipated supply of public rental housing. However, as some professional bodies and advocacy groups warned, these deregulation measures were followed by a period of rising rents in the private sector, beginning from the late 2000s. This rise has resulted in severe rental affordability problems, especially in the low-end of the private rental market, which accommodates a significant proportion of low-income renters. Indeed, speculative landlords have grasped the opportunity to make bigger profits by sub-dividing their rental flats into smaller units, which are known locally as 'sub-divided units' (SDUs). As the name indicates, these dwellings have been created through sub-division of existing private residential flats into two or more smaller units for rental purposes.

The proliferation of SDUs intensified around the late 2000s and early 2010s. According to official research conducted by the government (Census and Statistics Department, 2015, 2016), there were 85,500 households (195,500 persons) living in SDUs in 2014, which increased to 87,600 households (199,900 persons) in 2015. Some of these SDUs have independent toilets and cooking areas, while some resemble those that were common in the post-war years, namely partitioned rooms or bed-spaces with shared toilets and kitchens. SDUs are typically overcrowded, albeit more affordable than whole flats. Nevertheless, if we examine more recent data from the government's Thematic Household Survey, it becomes clearer that the so-called 'affordable' SDUs are actually not as affordable as they seem to be. According to the latest official data (Census and Statistics Department, 2015, 2016), the rent-to-income ratio of households living in SDUs was 30.8 per cent in 2014, which increased to 32.3 per cent in 2015. This increase was perhaps expected, since the 'newer' types of SDUs (with independent toilets and cooking areas) provide more privacy and thus are able to fetch higher rents than conventional rooms or bed-spaces. Although the median rent-to-income ratio of these SDUs is below 40 per cent, it still represents a substantial rental burden, especially for low-income households of bigger sizes, who need to cope with relatively high non-housing expenditures.

More importantly, discussions of rental affordability are meaningful only when we take into account the factor of housing quality. Housing which is 'affordable', but which is of low quality, indicates the existence of inadequate housing, which deserves public policy attention. Indeed, housing quality of SDUs in Hong Kong is considerably low: the average area per capita for SDU households was as low as 5.8 square metres in 2015 (Census and Statistics Department, 2016), compared to the average living space per capita of 13.1 square metres for public housing tenants in 2015 (Hong Kong Housing Authority, 2015). These figures suggest that, by the housing space standards alone, both housing affordability and housing quality in the low-end private rental sector have worsened in recent years.

The removal of rent control and security of tenure have certainly played parts in aggravating housing security and housing affordability problems in the private rental sector. Furthermore, these problems can be explained by rapid increases in private domestic rents in recent years. From 2006 to 2016, the rental index of private domestic flats increased from 91.6 to 168.2 (Rating and Valuation Department, 2017), representing a 84 per cent increase over a period of ten years. Rapid rent increases can possibly be explained by affordability problems in the owner-occupation sector, especially for first-time buyers, which has led to stronger demand for rental housing from those who are unable to afford home ownership.

In particular, rents have increased rapidly for small flats, which are known locally as Class A flats (flats with saleable area under 40 square metres). The rental index of Class A flats increased from 90.1 in 2006 to 137.1 in 2011, representing a 52 per cent increase; by 2016, the index reached 184.8, meaning that there was a 105 per cent increase over ten years (Rating and Valuation Department, 2017), which is a bigger increase than the overall increase for all domestic flats. This jump suggests that rent increases have pushed up demand for small rental flats, which subsequently lead to their rental increases, further driving the demand for ever-smaller sub-divided flats.

Socio-economic factors: immigration and income inequality

There are also other socio-economic factors that help explain trends observed in the private rental sector. One of the popular explanations that have been put forward for the growth in private rental households is the growth in immigrant households from Mainland China. As discussed above, the immigrant households are not eligible for public rental housing until they fulfil the residency requirements. If these immigrant households are unable to afford home ownership, their housing needs would have to be met through private renting, thus leading to rental increases. Government data appear to provide a certain degree of support for this argument. From 1996 to 2001, the number of domestic households with persons from the mainland having resided in Hong Kong for less than 7 years (PMRs) increased from 103,784 to 145,702, representing an increase of 40 per cent (Census and Statistics Department, 2006). From 2001 to 2006, the growth in households with PMRs was not as big, but there was still a moderate increase from 145,702 to 148,617 households (Census and Statistics Department, 2012a). Thus, growth in the number of immigrant households could partially explain the increase in demand for private rental accommodation from the late 1990s to the mid-2000s.

The median household income of households with PMRs has also been consistently lower than that of all households in Hong Kong. In 2001, the median household income of households with PMRs was HK$12,050, compared to HK$18,710 for all households; in 2006, the median household income of PMR households had dropped to HK$12,000, compared to HK$17,250 for all households. This discrepancy is not surprising, since immigrant households in Hong Kong are more likely to take up low-skilled jobs, such as service jobs and elementary occupations. Furthermore, the data from 2006 indicates that the majority of households with PMRs were concentrated in the Sham Shui Po district (Census and Statistics Department, 2006), which is an inner urban district characterized by a high concentration of SDUs (SoCO, 2013). Thus, this body of data renders further support for the argument that the growth in immigrant households has stimulated demand for accommodation in the private rental sector. However, this explanation seems to be less satisfactory for the period after the mid-2000s, since the number of immigrant households with PMRs actually declined from 148,617 households in 2006 to 115,323 in 2011 (Census and Statistics Department, 2012a). The percentage share of households with PMRs among all households in the whole territory declined accordingly, from 6.7 per cent to 4.9 per cent. Thus, we cannot attribute the growing demand for SDUs to immigration trends alone.

Another possible factor explaining the increasing demand for small rental flats is stagnant wage growth. As highlighted by Yip & La Grange (2006), the Asian Financial Crisis of 1997 resulted in cost-cutting measures in both the public and private sectors, including downsizing, pay freezes and wage cuts (p. 999). Wage increases have remained minimal, however, even more than ten years after the financial crisis, especially for the lower-income population.

According to data from the Census and Statistics Department (Census and Statistics Department, 2012b), the median monthly income from main employment for the working population[4] in the first (lowest) income decile was HK$3,500 in 2001, which dropped to HK$3,000 in 2006 and rose back to HK$3,580 only in 2011. In other words, there was only a 2 per cent increase over a period of ten years. For those in the second income decile, there was a 15 per cent increase from 2001 to 2011, while for those in the third income decile, there was a 6 per cent increase over these ten years.

These figures stand in contrast to the working population belonging to the higher income deciles. For those in the tenth (highest) income decile, median monthly income increased from HK$45,000 in 2001 to HK$55,000 in 2011 (a 22 per cent increase over ten years). Similarly, for those in the ninth income decile, there was a 20 per cent increase in the ten-year period from 2001 to 2011.

The absolute number of low-income households has also increased: the number of households with household income less than HK$6,000 increased by 82,678 households from 2001 to 2011, representing a 35 per cent increase, which is disproportionately higher than the 15 per cent increase in the overall number of households in the territory (Census and Statistics Department, 2012b). This skew implies that low-income households not only suffered from slow wage increases, but also expanded over the years, implying a higher demand for poor-quality rental housing in the private sector, especially for those households who are not eligible for public housing.

Slow wage increases among low-income workers can be linked to the changing occupational structure in Hong Kong, which has resulted in widening income inequalities and substantial differences in earnings across different industries (Chiu & Lui, 2004). Low earnings in low-skilled occupations have also been linked to the growth in non-regular, part-time and temporary workers (Yip & La Grange, 2006). It appears that employment conditions for low-pay, low-skilled workers have not improved much throughout the 2000s, which helps explain worsening affordability problems for low-income households and thus growing demand for SDUs as a more affordable option.

Overall, although the government has continued to build public rental housing, the supply of new public housing has not been quick enough to meet growing demand from the low-income population, which has been triggered by growing numbers of immigrant households and growing numbers of low-income households whose spending power has been constrained by slow wage growth and increasing rents in the private sector.

The future of private renting in Hong Kong

A key question that emerges from the above analysis is the possible role of the government in improving rental affordability and housing quality in the private rental sector in Hong Kong. Despite the continued surge in rents and emerging stories of hardships faced by displaced tenants, successive administrations have showed no intention of amending existing legislation to re-introduce rent regulation measures. So far, the government has provided no alternative measures to alleviate housing security and affordability problems in the private rental sector. At most, the government has proposed some piecemeal solutions, such as launching a pilot scheme in 2017 to encourage non-governmental organisations and social enterprises to operate so-called 'government-approved' subdivided rental flats. These subdivided flats are envisaged to be more liveable than existing SDUs, since they must comply with safety and sanitary standards. Yet even the government acknowledges that this is a small-scale and short-term initiative, with minimal impact on the overall problem.

The only long-term solution that has been put forward is increasing the production of public rental housing. However, experience has shown that there are limits to speeding up the supply of public housing. In particular, insufficient land for public housing construction and lengthy disputes in public housing projects (Lau, 2017) have posed major challenges to the government's housing supply strategy. If this situation continues, it would be increasingly unconvincing to view the private rental sector as a transitional sector only. Many households may continue to be stuck in the low-end rental market for years, and some may even become long-term private renters, a trend that has been already been recognized in other places such as Australia (Wulff & Maher, 1998).

Advocacy groups have tried to seize opportunities provided by the review of the Long-Term Housing Strategy in 2013 to lobby for rent control, security of tenure and in-cash rent subsidies for private tenants waiting for public housing. Yet the government has tended to reiterate the possible negative effects of these measures, such as the possibility that rent subsidies may push up rents in a tight market. Indeed, some lobbying groups have also acknowledged the potential disbenefits of re-introducing rent control under the present market conditions, especially the possible effect of reducing supply of rental properties, which may unintentionally exacerbate the imbalance between rental supply and demand.

Difficulties in achieving consensus on regulating the private rental sector are, as a matter of fact, not unique to Hong Kong. Contradictory views on whether or how to regulate the private rental sector have also existed in England, as discussed by Rugg & Rhodes (2003). In particular, there has been a lack of clear evidence on the relationship between regulation and rental property supply – such uncertainty regarding causes and effects makes it almost impossible to arrive at a consensus about the extension of regulatory measures. Instead of focusing on conventional regulatory measures such as rent control, there are alternative strategies such as fine-tuning existing legislation and providing adequate resources to enforcement agencies to speed up enforcement of existing statutory regulations to ensure fire and construction safety. In the long run, incentives or accreditation schemes could also be provided to encourage self-regulation by landlords and letting agencies (Rugg & Rhodes 2003, pp. 938–940).

Nevertheless, implementation of these alternative regulatory measures in Hong Kong involves considerable challenges. Local lobbying groups have already pointed out the problematic effects of strengthening enforcement against illegal SDUs, which may lead to displacement of low-income renters, who would have to compete among themselves for a shrinking supply of low-end rental flats. It appears that fine-tuning of regulatory measures would have to go hand in hand with an ongoing search for land to provide new public housing, in order to reduce private rental housing demand, improve rental affordability and housing quality problems in tandem. Regardless of the above challenges, it is becoming increasingly clear that the private rental sector is not the short-term, transitional sector that it is often imagined to be. As such, there is an urgent need for deliberation on long-term policies to improve housing wellbeing in this sector. In the absence of any explicit policy interventions, it is likely that the gap between the housing wellbeing of public renters and that of private renters will remain substantial in the coming years.

Conclusion

Since the mid-1990s, the governments of Hong Kong have attempted intermittently to promote home ownership as the major tenure with various subsidized schemes. These attempts have failed because of unfavourable economic conditions, housing market cycles, land development and supply policies. Thus, an asset-based welfare policy through housing has been impossible though aspired to. In contrast, the economic, political and social functions of PRH combined with the path-dependent nature of housing policy has made subsidized rental housing the core and perennial feature of Hong Kong's housing policy. This chapter views such a housing policy as a security-based welfare policy realized through housing. "Security-based welfare policy" refers to the security of tenancy and affordable housing. In the case of Hong Kong, it further refers to the security of housing-related facilities and services, plus facilitating the social wellbeing of the lower income groups. Most importantly, it fills a gap in the welfare system as the policy provides post-retirement protection by ensuring lower-income families affordable housing with its low-rent housing and rental assistance schemes. The welfare benefits of such a housing policy are similar to those offered by an asset-based welfare policy, not in terms of helping households to accumulate asset and climbing up the housing ladder (Kemeny, 1992, 2005), but in terms of housing security, affordability and social wellbeing. Over the years, PRH has become the cornerstone of Hong Kong's social stability. However, this stability has been debased in recent years due to social changes, notably demographic development, housing culture evolution and a new wage policy. Economic factors are at play as well: housing market conditions and repercussions of past land supply policy, which aggravate housing affordability and cause greater imbalance in the supply and demand of PRH.

The expanding private rental sector with deteriorating housing quality has reflected the limits of a security-based welfare policy embedded in a producer-led and rental-tenure-centred housing subsidy policy. The solution to such limits has been entrusted with increasing the land supply, but the question that remains unresolved is the lack of a deep and comprehensive understanding of the root causes of the increasing demand for PRH, and therefore effective solutions may still be remote. Is land shortage the only major reason for the rapid lengthening of the PRH waiting list? Why are the young people so keen or why do they feel the urgent need to apply for PRH? What are the root causes for the huge deterioration in housing affordability; is the trend transient and is the global trend of neo-liberalism to blame? How would ageing affect the nature and the quantum of housing demand? These are some of the questions that need in-depth investigation before the fundamental problems of the private rental market under a security-based housing policy can be tackled and the gap between the housing wellbeing of public and private rental housing can be narrowed.

Acknowledgement

The work described in this chapter was funded by grants from the Research Grants Council of the Hong Kong Special Administrative Region, China (Project no. HKU 742811H and Project no. HKU 17612215).

Notes

1 Housing type in Hong Kong is conventionally categorized according to the source of housing and not ownership, that is, accommodation provided by the government and subsidized NGOs is public housing, and that provided by the market is private housing.
2 Immigrants from mainland China are subject to a daily quota of 150 and constitute the largest source of population growth, e.g. contributing to four-fifths of net population growth in 2011–2015 (Research Office, Legislative Council Secretariat, 2017).
3 1 HKD = 0.13 USD (as of November 2017).
4 The data excludes Foreign Domestic Workers (FDWs), since FDWs in Hong Kong tend to live in their employers' accommodations and do not need to rent or buy a home.

References

Bradbrook, A. J. (1977). The future of domestic rent control in Hong Kong. *Hong Kong Law Journal*, 7, 321–361.

Castells, M., Goh, L. & Kwok, R. Y. W. (1990). *The Shek Kip Mei Syndrome: Economic development and housing policy in Hong Kong and Singapore*. London: Pion.

Cattell, V., Dines, N., Gesler, W. & Curtis, S. (2008). Mingling, observing, and lingering: Everyday public spaces and their implications for well-being and social relations. *Health & Place*, *14*(3), 544–561.

Census and Statistics Department. (1992). *Hong Kong 1991 Population census: Main tables*. Hong Kong: Census and Statistics Department.

Census and Statistics Department. (2006). *2006 Population by-census thematic report: Persons from the mainland having resided in Hong Kong for less than 7 years*. Hong Kong: Census and Statistics Department.

Census and Statistics Department. (2012a). *2011 Population census thematic report: Persons from the mainland having resided in Hong Kong for less than 7 years*. Hong Kong: Census and Statistics Department.

Census and Statistics Department. (2012b). *2011 Population Census thematic report: Household income distribution in Hong Kong*. Hong Kong: Census and Statistics Department.

Census and Statistics Department. (2013). *Older persons living in domestic households by living arrangement and type of housing, 2011*. Hong Kong: Census and Statistics Department.

Census and Statistics Department. (2015). *Thematic household survey report No. 57: Housing conditions of sub-divided units in Hong Kong*. Hong Kong: Census and Statistics Department.

Census and Statistics Department. (2016). *Thematic household survey report No. 60: Housing conditions of sub-divided units in Hong Kong*. Hong Kong: Census and Statistics Department.

Census and Statistics Department. (2017a). *Hong Kong 2016 population by-census – thematic report: Household income distribution in Hong Kong*. Hong Kong: Census and Statistics Department.

Census and Statistics Department. (2017b). *Median rent to income ratios by type of quarters and year*. Hong Kong: Hong Kong Special Administrative Region.

Chief Secretary's Administration Office. (2015). *Population policy: Strategies and initiatives*. Hong Kong: Chief Secretary's Administration Office.

Chiu, R. L. H. (2001). The role of the government in housing in socialist China and capitalist Hong Kong. *Third World Planning Review*, *23*(1), 5–21.

Chiu, R. L. H. (2003). Analysis on policy trends. In Y. M. Yeung & K. Y. Wong (Eds.), *Fifty years of public housing in Hong Kong: A golden jubilee anniversary public housing volume*. Hong Kong: Chinese University Press, pp. 221–242 (In Chinese), pp. 235–258 (In English).

Chiu, R. L. H. (2005). Housing in the Social Development Perspective. In R. J. Estes (Ed.), *Social development in Hong Kong: The unfinished agenda*. London and New York: Oxford University Press, pp. 157–169.

Chiu, S. W. K. & Lui, T. L. (2004). Testing the global city-social polarization thesis: Hong Kong since the 1990s. *Urban Studies*, *41*(10), 1863–1888.

Costa-Font, J. (2013). Housing-related well-being in older people: The impact of environmental and financial influences. *Urban Studies*, 50(4), 657–673.

Cox, K. R. (1982). Housing tenure and neighborhood activism. *Urban Affairs Quarterly*, 8, 107–129.

Diaz-Serrano, L. (2009). Disentangling the housing satisfaction puzzle: Does homeownership really matter? *Journal of Economic Psychology*, 30(5), 745–755.

Forrest, R., La Grange, A. & Yip, N. M. (2002). Neighbourhood in a high rise, high density city: some observations on contemporary Hong Kong. *Sociological Review*, 50(2), 215–240.

HKSAR Government. (2003a). *Consultation paper: Landlord and tenant (consolidation) ordinance (lto) (cap. 7) security of tenure*. Hong Kong: Housing, Planning and Lands Bureau.

HKSAR Government. (2003b). *Landlord and tenant (consolidation) ordinance (cap. 7) security of tenure and related provisions: Result of consultation exercise*. Hong Kong: Housing, Planning and Lands Bureau.

Hong Kong Housing Authority. (1993). *A report on the mid-term review of the long term housing strategy*. Hong Kong: Hong Kong Housing Authority.
Hong Kong Housing Authority. (2006). *Waiting list income and asset limits*. Hong Kong: Hong Kong Housing Authority.
Hong Kong Housing Authority. (2015). *Housing in figures: 2015*. Hong Kong: Hong Kong Housing Authority.
Hong Kong Housing Authority. (2016). *2015/16 Annual report: Caring for people, committed to progress*. Hong Kong: Hong Kong Housing Authority.
Hong Kong Housing Department. (2010). Design for a green & healthy living in public rental housing estate in Hong Kong. *Paper presented to the 22nd IFPRA World Congress, November 2010*. Hong Kong: Leisure and Cultural Services Department, Hong Kong Recreation Management Association.
Hong Kong Institute of Architects. (2012). *Hong Kong today: Public housing – improvement in quality of life*. Hong Kong: Hong Kong Institute of Architects.
Hu, F. (2013). Homeownership and subjective wellbeing in Urban China: Does owning a house make you happier? *Social Indicators Research*, *110*(3), 951–971.
Hulse, K. & Milligan, V. (2014). Secure occupancy: A new framework for analysing security in rental housing. *Housing Studies*, *29*(5), 638–656.
Kemeny, J. (1992). *Housing and Social Theory*. London: Routledge.
Kemeny, J. (2005). "The really big trade-off" between home ownership and welfare: Castles' evaluation of the 1980 thesis, and a reformulation 25 years on. *Housing, Theory and Society*, *22*(2), 59–75.
Lau, M. (2017). Framing processes in planning disputes: Analysing dynamics of contention in a housing project in Hong Kong. *Housing Studies*. Available at https://doi.org/10.1080/02673037.2017.1383367 (published online October 2017), accessed on 13 November 2017.
Lawton, M. P., Nahemow, L. & Teaff, J. (1975). Housing characteristics and the well-being of elderly tenants in federally assisted housing. *Journal of Gerontology*, *30*(5), 601–607.
Legislative Council. (2004, June 30). *Minutes of the Hong Kong Legislative Council meeting*. Hong Kong: Legislative Council Secretariat.
Phillips, D. R., Siu, O. L., Yeh, A. G. O. & Cheng, K. H. C. (2005). The impacts of dwelling conditions on older persons' psychological well-being in Hong Kong: The mediating role of residential satisfaction. *Social Science & Medicine*, *60*(12), 2785–2797.
Rating and Valuation Department. (2017). *Property market statistics. Private domestic – rental indices by class (territory-wide)*. Hong Kong: Rating and Valuation Department.
Research Office, Legislative Council Secretariat. (2017). *Statistical highlights issue ISSH20/16–17*. Hong Kong: The Legislative Council Commission.
Rugg, J. & Rhodes, D. (2003). 'Between a rock and a hard place': The failure to agree on regulation for the private rented sector in England. *Housing Studies*, *18*(6), 937–946.
Society for Community Organization (SoCO). (2013). *2012/13 Research report on cage homes, cubicles, and sub-divided flats*. Hong Kong: Society for Community Organization.
The Government Information Centre. (2011, 2016). *Waiting list income and asset limits*. Hong Kong: The Government Information Centre.
Transport and Housing Bureau. (2014). *Long-term housing strategy December 2014*. Hong Kong: Government Logistic Department.

Tung, C. H. (1997, July 1). *A speech by the Chief Executive: A future of excellence and prosperity for all.* Available at www.info.gov.hk/isd/speech/0701ceho.htm, accessed on 29 November 2017.

Wulff, M. & Maher, C. (1998). Long-term renters in the Australian housing market. *Housing Studies*, *13*(1), 83–98.

Yip, N. M. & La Grange, A. (2006). Globalisation, de-industrialization and Hong Kong's private rental sector. *Habitat International*, 30(4), 996–1006.

4 Housing in Japan's post-growth society

Yosuke Hirayama and Misa Izuhara

Introduction

The housing system in Japan during the 'post-war growth period' was focused on driving the mass construction of owner-occupied housing (Hirayama, 2007; Oizumi, 2007). This effort revolved around various 'upward' social changes, including increasing population, rapid economic development and rising real incomes. The government placed emphasis on promoting middle-class home ownership in formulating housing policy, providing many households with financial assistance for building or purchasing their own homes. There was a robust cycle in which the mass production of owner-occupied housing accelerated economic growth, and this growth, in turn, further encouraged the acquisition of owner-occupied houses. In many countries, housing systems have not only provided shelters but have also been embedded in broader social processes (Forrest & Lee, 2003). Underlying Japan's post-war development was a housing system oriented towards expanding housing construction and facilitating middle-class home ownership.

Since the 1990s, however, Japan has entered a 'post-growth era' along with demographic and economic shifts to stagnation, which has radically reorganized the condition of the housing system (Hirayama & Izuhara, 2018). The nation has undergone continuing economic decline since the bubble economy collapsed at the beginning of the 1990s. Many renters have faced increasing difficulties to enter the owner-occupied housing market while the sustained fall of land and housing prices has undermined the security of housing as an asset. The total population of Japan, which began to decline in the mid-2000s due to extraordinarily low fertility, will continue to decrease with an accelerated rise in the proportion of elderly people. This shift will inevitably encourage a decline in investment in housing. Moreover, within the context of pervasive neoliberalism, the Japanese government has retreated from various housing subsidy schemes to move towards accentuating the role of the market in producing and consuming housing. Thus, people's fortunes related to housing in the post-growth era are now completely different from those in the post-war high-rate growth period.

During the growth era, the aggregated flow of people entering the home ownership sector underpinned the construction and maintenance of a society that was centered on the middle classes. With the transition into the post-growth era, however, changes in housing conditions have led to the expansion of social inequalities both between and within generations. Many older cohort households entered the owner-occupied housing sector and accumulated property assets during the period of high-speed growth, whereas younger cohort households have increasingly found it difficult to access home ownership with continuing stagnation in the post-growth era. Within younger generations, there have been widening gaps, in terms of the ability to secure adequate housing, between those having stable employment and the increasing number of those with more precarious employment. Among older generations, property ownership has increasingly become stratified. Elderly households with higher-incomes own houses with higher property values, while those with lower-incomes tend to live in owner-occupied housing with lower property values.

In this chapter, we look at changes in Japan's housing conditions within the post-growth context. Many mature economies have begun to experience post-growth social changes characterized by slower economic growth, aging populations, lower fertility, and increasing social inequality. Japan stands at the leading edge of transitions into a post-growth society in the East Asian region as well as the Global North, providing a vivid case study with regard to housing developments after the growth age. We begin by identifying key elements such as the demographic, economic and policy shifts that shape post-growth social transformations in Japan. This is followed by various analyses of housing conditions in the post-growth age. We then move on to exploring the role that housing plays in reorganizing inter- and intra-generational social inequalities. Thus, the overall aim of this chapter is, using Japan as a case study, to highlight the importance of the post-growth contexts when investigating contemporary housing and social processes in housing systems with declining or slow growth economies and aging populations.

Towards a post-growth society

Demographic development

We focus on three key ingredients that constitute post-growth social transitions for our exploration of housing developments, namely, shrinking and aging population, continuing economic stagnation and policy changes towards a more neoliberal direction. First and foremost, demographic decline is one of the most important elements responsible for Japan's shift to a post-growth society. According to the 2012 estimate by the National Institute of Population and Social Security Research, the population, which was 128 million in 2010, will decrease to 107 million in 2040 and then to 87 million by 2060. This decrease will occur because Japan has one of the lowest fertility rates in the world. The total fertility rate (TFR) in Japan, which was more than 2.0 until the mid-1970s, decreased to a record low of 1.26 in 2005. The rate has since increased slightly but has remained at a very low level: for example, 1.39 in 2010 and 1.42

in 2014. The population decrease is being paralleled by a rapid increase in the proportion of elderly people. Japan became a 'super-aged society' in 2007, when the percentage of those aged 65 or more exceeded 21 per cent. This figure is forecast to increase to as much as 36.1 per cent in 2040 and 39.9 per cent by 2060. The proportion of elderly people in Japan has been, and will continue to be, the highest in the world.

Such demographic changes have a significant impact on the condition of the housing system. The government has traditionally encouraged the mass construction of dwellings as a way of stimulating economic expansion. Pressure from the demand for housing will be, however, significantly lessened in the future due to the shrinking and aging population. It will thus become more difficult to maintain a high level of investment in housing production. If the number of households continues to increase with a decrease in household size, it might be possible, despite the population decrease, to maintain high levels of new housing starts. However, the number of households is expected to begin decreasing around 2020.

In addition, the shift to a post-growth society has been accompanied by a decline in the number of marriages and associated changes in household patterns. In post-war Japan, acquiring one's own home was closely linked with establishing a family. Not only married couples but also singles and cohabiting couples aspire to home ownership in some Western countries, whereas in Japan most people do not purchase a dwelling until they marry. The government encouraged family households to access home ownership in terms of implementing housing policy, while the tax, social security and corporate-based welfare systems gave a range of economic advantages to family households. The combination of marriage and home ownership was indeed at the heart of Japan's post-war social development. Therefore, the decline in marriage has meant a fundamental challenge to the traditional housing system oriented towards home ownership.

There has been a notable increase in those who either delay marriage or never marry for various reasons. According to the Population Census, between 1980 and 2010, the percentage of unmarried people in the 30 to 34 age group rose from 21.5 per cent to 47.3 per cent for men and from 9.1 per cent to 34.5 per cent for women. Unmarried rates are much higher for males than for females. Moreover, the ratio of never-married individuals at 50, who are defined as 'lifelong never-married persons' in Japanese statistics, increased to 20.1 per cent for men and 10.6 per cent for women by 2010. This figure is, according to the 2013 estimate presented by the National Institute of Population and Social Security Research, expected to increase to 29.0 per cent for men and 19.2 per cent for women by 2035. With a decrease in marriages, household patterns have changed. Between 1980 and 2010, the percentage of one-person households increased from 15.8 per cent to 32.4 per cent, while that of households comprised of married couples and children decreased from 44.2 per cent to 27.9 per cent.

Continued economic stagnation

The economic bubble in Japan, which swelled in the latter half of the 1980s with an abnormal upsurge in real estate prices, collapsed at the beginning of the 1990s. This collapse marked a turning point in which Japan entered a new era of post-growth stagnation. Annual real growth rates in GDP on average, which were more than 5 per cent in the 1970s, fell to less than 4 per cent in the 1980s, to less than 2 per cent in the 1990s, and then to less than 1 per cent in the 2000s. The post-bubble recession was extremely prolonged and traumatic, with minimal or negative real growth in GDP (Oizumi, 1994). The banking sector was plunged into crisis due to a huge number of unrecoverable loans, provoking the 1994 and 1997 financial crises with chain-reaction bankruptcies of major banks and security firms as well as hardship in other sectors. Real incomes were reduced, and the employment market was destabilized. The Japanese economy eventually began to recover in 2002. However, the economic upturn did not translate into an improved household economy, and Japan entered a recession again in 2008, having become entangled in the global financial crisis. Subsequently, the European debt crisis began to affect many economies including Japan in 2009. Since the early 2010s, the Japanese government has accelerated financial deregulation to escape from prolonged stagnation, and the Bank of Japan has implemented unprecedented monetary easing. Despite these unconventional financial measures, the economy has not yet recovered, but rather has remained obstinately stagnant. Thus, since the early 1990s, economic insecurity has come to be perceived as 'normal' rather than 'abnormal'.

In addition, demographic changes have been playing a role in the economic downturn. The population decrease combined with an increase in the proportion of elderly people means a substantial decline in those who are of productive age. The percentage of those aged between 15 and 64, which was 63.8 per cent in 2010, is projected to decrease to 53.9 per cent in 2040 and to 50.9 per cent by 2060. Such a substantial decrease in the workforce will presumably constitute a more significant barrier against economic recovery.

Economic decline has led to the rapid shift of the labor market to 'casualization'. Japan's post-war development was characterized by the formation of a 'company society', where major corporations adopted a life-long employment system and a seniority system for wages and promotion (Fujita & Shionoya, 1997; Shinkawa, 1993). However, within the context of the post-bubble economic downturn, and in response to the intensified competitiveness of the globalizing business environment, an increasing number of corporations, struggling for their survival, carried out large-scale personnel restructuring and downsizing. This restructuring weakened the foundation of the 'company as a family' model. The labor market has thus been reoriented around declining employment stability with an associated rapid increase in non-regular, low-wage employees. According to the Employment Status Survey, the number of regular employees decreased from 38.1 million to 32.8 million between 1994 and 2014, while the number of non-regular employees, including dispatched workers, short-term contract-based employees, part-time workers, and temporary employees increased sharply from 9.7 million to 19.6 million in the same period. The percentage of non-regular employees versus all employees rose to 37.4 per cent in 2014.

During the post-war growth period, many households with a rise in real incomes acquired owner-occupied homes and accumulated valuable real estate assets. Since the bubble burst, however, prolonged stagnation and associated changes in the labor market have fundamentally affected the circumstances surrounding the housing system. As will be discussed later, a decline in real incomes has made it more difficult for many households to secure affordable housing while at the same time there has been serious devaluation in residential properties. An increase in the number of workers in precarious employment circumstances has meant that some workers are now experiencing difficulties in finding places to live (Iwata, 2009).

Housing policy reorientation

The Japanese government during the post-war growth period energetically intervened in the housing market, encouraging more middle-class households to acquire owner-occupied housing. Promoting home ownership was considered instrumental in facilitating economic expansion and maintaining social stability. The Government Housing Loan Corporation (GHLC), founded as a state agency in 1950, provided many households with long-term, fixed-low-interest mortgages for purchasing or constructing houses. This meant that the corporation formed a separate and privileged circuit of housing loans and helped to alleviate the influence of the volatile financial market on borrowers. The government sought to create a post-war society centered on middle-class property ownership, placing particular emphasis on protecting home owners. According to the Housing and Land Survey, since the 1960s, despite rapid urbanization, with the provision of GHLC loans aimed at encouraging housing acquisition, the level of owner-occupation has stayed at approximately 60 per cent (for example, 61.1 per cent in 2008 and 61.7 per cent in 2013), representing its position as the dominant housing tenure.

Meanwhile, the government did not significantly seek to improve the conditions of the rented housing sector. The ratio of private rented housing has been considerably high, accounting for more than a quarter of all housing (28 per cent) in 2013. However, the government has not supported the supply of private rented housing. There has been little assistance for the construction of private rental housing and almost no provision of rental subsidies. Public rented housing has been allocated to low-income households, while the Japan Housing Corporation, which was founded as a state agency in 1955, and local public corporations have constructed rented housing for urban middle-income households. The direct supply of rented housing by the public sector has, however, been positioned as a residual measure. The proportions of low-income public housing and rented housing of public corporations in the total housing stock have been low, e.g., 3.8 per cent and 1.6 per cent, respectively, in 2013.

Entering into the post-growth period, the government reorganized housing policy towards emphasizing market solutions in alignment with progressively prevalent neoliberal prescriptions. Since the early 1970s, when capitalist societies underwent economic crises such as the oil crisis and the collapse of the Bretton Woods system, declines in growth have stimulated the ascendance of neoliberal imperatives, which have been increasingly influential in reorientating state policies. The introduction of neoliberal policies was notably slower in Japan than in Britain or the United States, where policy shifts began to appear in the early 1980s (Forrest & Hirayama, 2009). This slowness occurred partly because Japan's economic performances were relatively strong even after the oil crisis. Nonetheless, the economic crisis following the bursting of the bubble in the early 1990s fueled a reorientation of state policies.

Since the mid-1990s, the Japanese housing system has undergone a radical reorganization in line with neoliberal prescriptions, emphasizing the role played by the market economy in providing housing and mortgages (Hirayama, 2007). The system of providing low-income public housing, which had traditionally been residual, was further marginalized. The Housing and Urban Development Corporation, the successor to the Japan Housing Corporation, was reorganized into the Urban Development Corporation in 1999 and again into the Urban Renaissance Agency in 2004. The new agency reduced its commitment to housing schemes substantially so as not to compete with the private sector. The GHLC was a central instrument in expanding middle-class home ownership. Nevertheless, the government began to reduce the provision of GHLC loans in the mid-1990s, and the corporation was ultimately abolished in 2007. The dissolution of the state agency, which had constituted the core of the housing system, was a particularly important watershed in the post-war history of Japan's housing policy. While the Housing Finance Agency was established as the successor to the GHLC, it withdrew from the primary mortgage market and thereafter dealt only with the secondary market of mortgage securities.

The rapid expansion of mortgages supplied by private banks filled the large void in the housing loan market created by the abolition of the GHLC. The housing loan market was deregulated in terms of interest rates in 1994, which led to a more competitive business environment for the mortgage market as well as improved lending conditions. However, unlike the GHLC, which provided fixed-low-interest loans, private banks provide mainly variable-interest mortgages, and home buyers, who take out variable-interest bank loans, are more directly exposed to fluctuations in the financial market. Furthermore, the GHLC provided all borrowers with mortgages under relatively homogeneous conditions in terms of interest rates and loan limits, whereas private banks supplied more diverse mortgage commodities under varied lending conditions, in particular, higher-interest loans to those with lower incomes. Neoliberal housing policy has thus been aimed at deregulating and expanding the housing loan market while undermining the system of protecting households accessing home ownership.

The shrinking of the housing and mortgage markets

How then have housing conditions transformed within the post-growth context? Changes in the circumstance of the housing system have been seen at the individual and household levels while simultaneously intertwining with broader social processes at large. Therefore, we should look at the housing situation at both the macro and micro levels. One of the most important features of changes at the macro level after the growth period is a decline in the volume of housing construction. As has been discussed, mass production of dwellings has underpinned Japan's post-war economic development. From the immediate post-war period to the high-speed growth period, there was an enormous housing shortage. This combined with the increase in the middle classes to accelerate the construction of housing. Sustained mass housing production, however, led to a reduction of housing shortage, and may even reverse the imbalance, causing housing surplus (Miyake, 1996). The number of existing dwellings in Japan, which was approximately 29 million in 1973, continued to increase to approximately 61 million in 2013, while the vacancy rate, which was 5.5 per cent in 1973, increased to 9.8 per cent in 1993, to 12.2 per cent in 2003, and to 13.5 per cent in 2013. As the housing market matured, housing starts inevitably decreased. Furthermore, upon Japan's entering the post-growth phase, demographic and economic stagnation began to shrink the housing market. The annualized average number of new starts, which was 1,507,152 in 1990–1994, significantly decreased to 1,180,770 in 2000–2004 and to 880,465 in 2010–2014 (see Figure 4.1). Between 1990 and 2014, the annual investment in housing decreased from 26.9 trillion to 13.9 trillion yen,[1] and its percentage versus real GDP dropped from 5.9 per cent to 2.6 per cent (see Figure 4.2).

In parallel with the decline in the housing market, the mortgage market has also begun to shrink. The total amount of outstanding housing loans for individuals increased dramatically from 45.1 billion yen in 1980 to 183.4 billion yen in 2000, and its percentage versus GDP rose from 18.6 per cent to 35.9 per cent during the same period. These figures have, however, been largely stationary since the early 2000s. This fact suggests that the mortgage market has matured and that with demographic and economic decline, it will become more difficult to maintain the market volume of housing loans. As can be seen in Figure 4.3, new issues of mortgages have begun to decline. The annualized average amount of newly issued housing loans for individuals decreased from 30.1 trillion yen in 1995–1999 to 20.0 trillion yen in 2010–2014.

In many developed economies including Japan, facilitating the supply of housing loans has been a key to stimulating economies (Aalbers, 2016). Since the beginning of the post-growth era, however, demographic and economic changes have had constraining effects on mortgage markets. The provision of housing loans in Japan began to increase in the 1970s, and the expansion continued until the mid-1990s, when the country entered the post-growth era, which underpinned the mass construction of housing and thus economic expansion. From the end of the war to the high-speed growth period, when banks

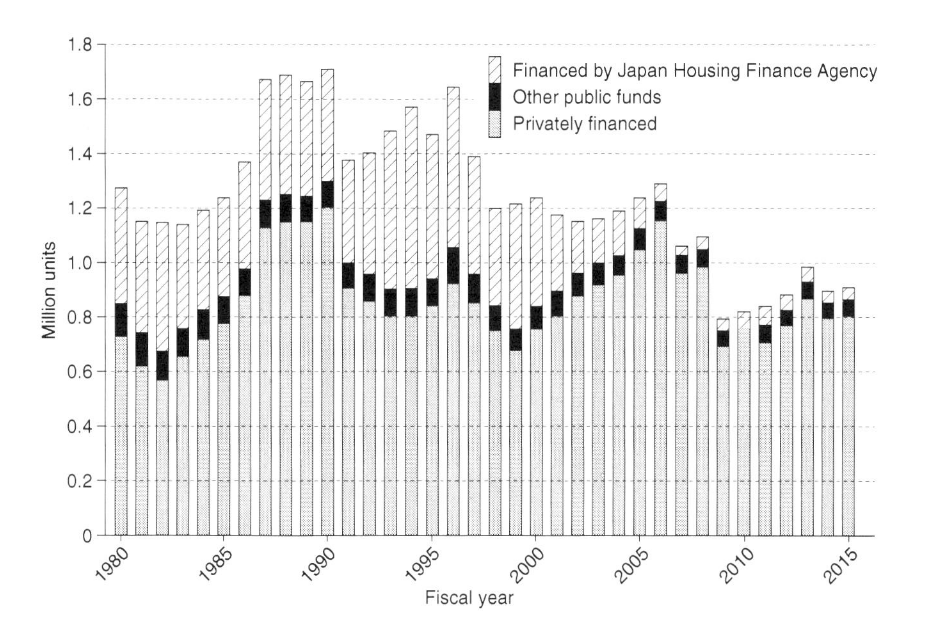

Figure 4.1 Housing new starts by type of fund.

Source: Ministry of Land, Infrastructure, Transport and Tourism.

Note

Japan Housing Finance Agency was formerly the Government Housing Loan Corporation.

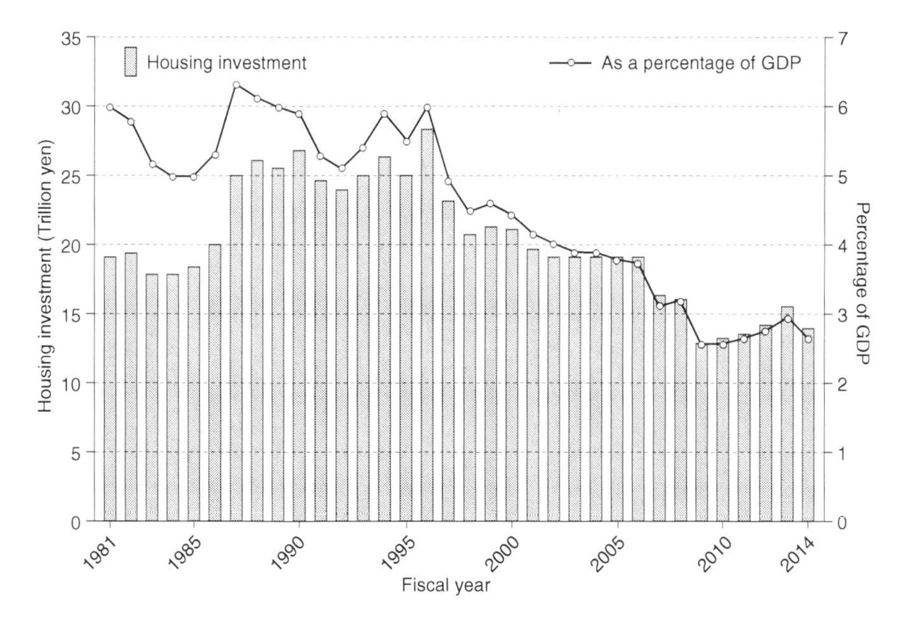

Figure 4.2 Housing investment and its percentage of GDP.

Source: Cabinet Office.

concentrated on financing investments in industrial development, the GHLC played a pivotal role in providing mortgages to those who wished to purchase their own housing. The banking sector, however, began increasing financing for real estate as big companies moved to obtain capital from equity finance in the 1980s, and subsequently, the government re-orientated public policies towards market solutions in line with neoliberal prescriptions in the 1990s. Consequently, the GHLC's loan system was replaced by bank loans in phases, leading to a privatization and marketization of residential mortgages (see Figure 4.3). Moreover, the government has promoted financial deregulation, which is supposed to expand the housing loan market.

Streeck (2014) argued that there has been a shift from public debt to private debt in neoliberal policy to cope with economic crises. This shift was best reflected in an increase in mortgaged housing. Households, rather than governments, were encouraged to take on debt to stimulate their countries' economies, which led to a noticeable increase in indebted home owners. This increase was embedded within the broader framework of a growth strategy that Crouch (2011) dubbed 'privatized Keynesianism'. Nevertheless, as has been argued, within the post-growth context, it is becoming progressively more difficult to expand mortgage markets. In Japan's post-growth ultra-aged society, new issues of housing loans have begun to decline despite financial deregulation. Many developed economies have undergone a shrinking of mortgage markets since the

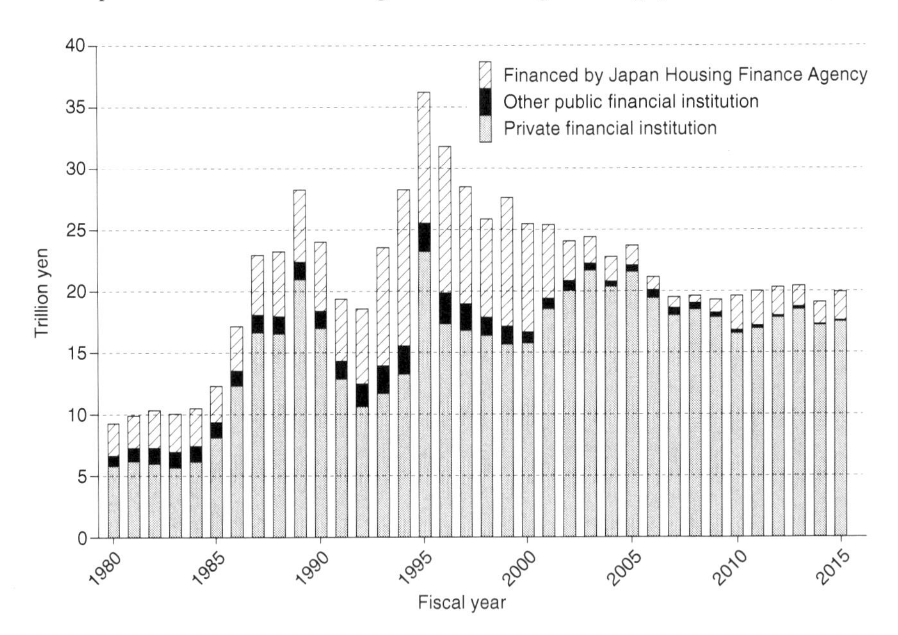

Figure 4.3 New issues of housing loans for individuals.

Source: Japan Housing Finance Agency.

Note

Japan Housing Finance Agency was formerly the Government Housing Loan Corporation.

global financial crisis. Thus, the housing situation in the emerging post-growth era raises questions as to the extent to which an economic strategy depending on debt financing for owner-occupied housing can be maintained.

Deflationary economy and housing affordability

Within the context of the transition to a post-growth society, a significant consequence of changes in housing conditions at the household level is a decline in housing affordability (Hirayama, 2010a). During the period of post-war high-speed economic growth, the inflationary economy accelerated a rise in housing costs, which led to a decrease in the affordability of housing. After the bubble burst, Japan entered a post-growth period associated with deflation or minimal inflation, and housing prices dropped. This drop might have been expected to improve affordability. But in reality, the economic burden of accessing housing did not decrease but rather increased, mainly due to deflation of real incomes. In other words, there has been a shift from house price inflation to income deflation with regard to the cause of the decline in housing affordability. In addition, the reorientation of housing policy towards emphasizing market solutions has combined with the post-bubble recession to undermine the affordability of housing. Japan's experience suggests that the end of the high-speed growth era and the beginning of the post-growth era do not necessarily alleviate housing affordability problems but rather restructure the mechanisms of determining the extent to which housing is affordable.

Post-bubble changes in the economic conditions of the rental housing sector have been characterized by an increase in lower-income households, a decrease in lower-rent housing, and a consequent increase in households with heavier burdens imposed by rent payments (Hirayama, 2010a). According to data on renter households obtained from the National Survey of Family Income and Expenditure, between 1994 and 2014, the nominal monthly disposable income decreased from 320,300 yen to 297,200 yen on average, whereas the average nominal monthly housing cost increased from 42,500 yen to 50,700 yen. As a result, the housing cost-to-disposable income rate of renter households increased from 13.3 per cent in 1994 to 17.1 per cent in 2014.

The stagnant economy and the associated decline in real incomes did not encourage a drop in the average rent. The main factor behind this is the reduced availability of low-rent housing. In Japan, the 'decommodified' low-rent housing sector, which is formed outside the market, consists of public rented housing and corporate-based employee housing. Neoliberal housing policy has, however, encouraged the further marginalization of public rented housing that has traditionally been residual, while the provision of employee housing has been decreasing, reflecting prolonged recession. Thus, there has been a continual decline in the decommodified rented housing sector (Hirayama, 2014). Low-rent housing has also been provided within the market domain. As the supply of decommodified rental dwellings has been limited in Japan, private rented housing has occupied the main position in the rented housing market. Although providing private rented housing has not been particularly profitable, a number of individuals or families who hold land have constructed and provided rental properties without investing in site acquisition. This practice has made the supply of low-rent housing possible. The rents of multifamily housing in wooden structures in particular have been set low. Although dwellings of this type are mostly substandard in terms of floor area and amenities, they function as low-cost accommodations for low-income households. The number of existing low-rent private dwellings has, however, decreased substantially as a result of structural aging or dilapidation (Hirayama, 2014).

The affordability of housing has declined not only in the rented housing sector but also in the home ownership sector (Hirayama, 2010a). Since the bubble collapsed, land and housing prices have largely been on the decline, and interest rates have decreased to an extraordinarily low level. Nevertheless, the economic conditions for purchasing a property have become more disadvantageous due to a decline in real incomes and an increase in mortgage repayments. Along with the prolonged post-bubble stagnation, the income decline has translated into higher loan-to-value ratio, i.e., smaller deposits and larger mortgage liabilities. As a result, despite the drop in real estate prices, the affordability of home ownership has not improved. An examination of the same survey (the National Survey of Family Income and Expenditure) as to the economic situations of home owners with mortgage liabilities reveals that between 1994 and 2014, the nominal monthly disposable income on average decreased from 485,600 yen to 424,700 yen, while the average nominal monthly housing cost, consisting mainly of mortgage repayments, increased from 63,900 yen to 78,500 yen. The proportion of disposable income spent on mortgage repayment therefore increased from 13.2 per cent in 1994 to 18.5 per cent in 2014.

The collapse of the bubble led to many households' having a harder time in procuring a mortgage. On the other hand, the government sought to accelerate acquisition of owner-occupied housing, with the aim of stimulating the economy. The supply of GHLC loans increased to an unprecedented level in the first half of the 1990s, and interest rates have remained low. This situation has encouraged housing purchasers to take out larger loans. The government re-orientated housing policy to neoliberal prescriptions in the mid-1990s, which led to the 1994 deregulation of the private mortgage market in regard to interest rates. The deregulation created a more competitive business circumstance for the housing loan market and resulted in improved lending conditions such as higher loan-to-value. Consequently, an increasing number of households have come to have larger outstanding mortgage debts despite income deflation.

Devaluation of residential properties

The decline in the security of home ownership as an asset base is another significant feature of the post-growth society at the household level. Since the bursting of the economic bubble, Japan has experienced a devaluation of residential properties, with a shift from capital gains to capital losses, resulting in a fundamental change in the home ownership economy. During the period of high-speed economic growth, many households aspired to home ownership because owning a house produced considerable capital gains and thus was a primary mechanism for accumulating assets. Renters entering the owner-occupied housing market as first-time buyers were able to expect an appreciation in the real value of properties while those owning a property had prospects of moving to a better house using the current property as a stepping stone. Capital gains indeed fueled the system of propelling people towards the top of the property ladder. Since the bubble burst, however, most residential properties have fallen in value.

Almost all households that have purchased a dwelling since the bubble period have experienced a devaluation of their property. The extent of housing depreciation has, however, varied between dwelling types; i.e., the rate at which housing prices have fallen has been greater in second-hand housing than in newly built housing and in condominiums rather than in single-family dwellings (Hirayama, 2011). The scale of capital losses from condominiums has thus been substantial. For example, the average price of a newly built condominium with a floor area of 70 square metres was 67.2 million yen in the Tokyo region in 1991. This average fell to 20.3 million yen by 2008, generating a capital loss of 46.9 million yen. In other words, the value of condominiums purchased during the peak of the bubble fell sharply, by some 70 per cent (Hirayama, 2011).

Institutional and policy factors have contributed to the sharper fall in the prices of second-hand condominiums. Japan's housing policy has focused on driving mass construction of dwellings as a means of stimulating the economy. In response to the post-bubble recession, the government expanded production of owner-occupied housing in line with the traditional strategy. This expansion led to over-construction of new housing and consequently accelerated a drop in the prices of second-hand housing. The government has emphasized urban regeneration as a key policy for large cities (Hirayama, 2017; Saito, 2012; Waley, 2015). This emphasis has facilitated the construction of many new condominium blocks, resulting in a notable decline in the prices of second-hand condominiums. Moreover, the GHLC provided borrowers who were purchasing new housing with loans under more favorable conditions than for those who were buying second-hand housing. The tax system also gave advantages to purchasers of new housing. Since the 2000s, the government has begun to place more weight on the expansion of second-hand housing transactions. This focus occurs due to changes in the condition of the housing market, including the beginning of the population decrease and the increase in vacant housing. However, the effect of this policy on strengthening the marketability of second-hand housing is slow in coming. This delay suggests that considerable time is required to expand the existing housing market, especially when one takes into account the long history of construction-oriented housing policy.

The housing asset conditions of home owners have been substantially undermined since the bubble burst. According to data on owner-occupier households with two or more members who hold mortgage liabilities (see Figure 4.4),

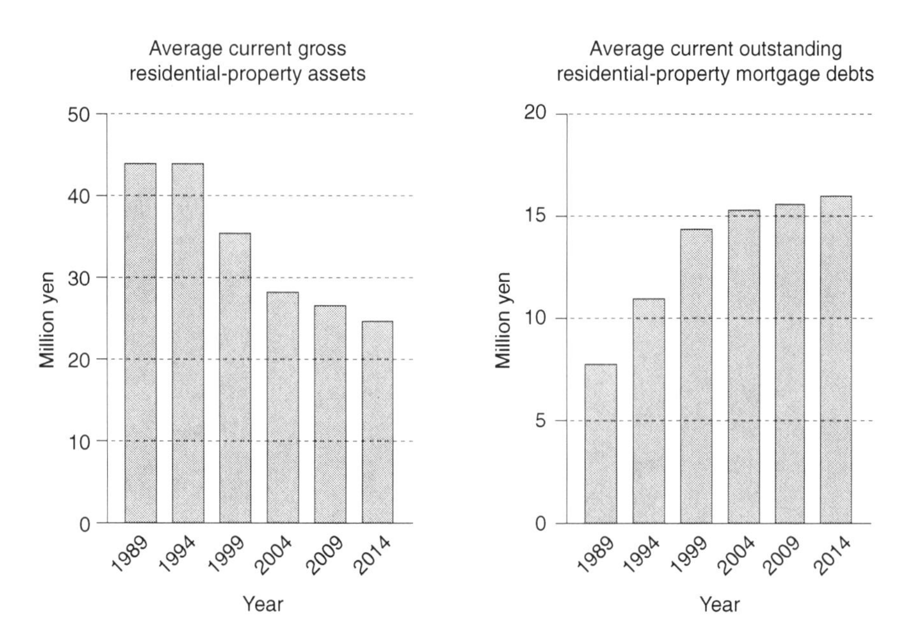

Figure 4.4 Changes in outstanding mortgage debts and gross assets on residential properties of owner-occupier households with mortgage debt.

Source: National Survey of Family Income and Expenditure, conducted by the Statistics Bureau.

Note
Data are for households with two or more members.

between 1994 and 2014, the average amount of current outstanding residential property mortgage debts increased from 10.9 million yen to 16.0 million yen, while that of current gross residential property assets decreased from 44.0 million yen to 24.5 million yen. This situation means that there has been a sharp decrease in the average amount of housing equity: from 33.1 million yen in 1994 to 8.5 million yen in 2014. There has thus been an increase in the number of home owners captured by negative equity. In practice, younger households who had purchased expensive housing at the peak of the bubble with high loan-to-value mortgages were confronted by negative housing equity. According to the National Survey of Family Income and Expenditure, from 1989 to 2004, of owner-occupier households with mortgage liabilities, those with negative housing equity increased from 5.9 per cent to 23.6 per cent for households with a head aged 34 or younger and from 2.7 per cent to 24.0 per cent for the 35-to-44-year-old group.

The insecure post-bubble economy translated into an increase in mortgage defaults, heralding the beginning of a new era characterized by higher risks with regard to home ownership (Hirayama, 2011). Foreclosures of residential properties with negative equity have brought about particularly problematic situations where former owners have lost their houses but continued to service remaining debt liabilities. As mortgage defaults increased, the GHLC was pressed to launch a series of schemes to assist borrowers who could not repay their housing loans. As a result, increasing numbers of borrowers were permitted to extend mortgage repayment periods. With the abolition of the GHLC and deregulation of the housing loan market in terms of interest rates, banks have provided increasingly diverse mortgage products with regards to interest rates and guarantee fees. This diversification has led households with lower-incomes to take out riskier housing loans, resulting in more defaults.

Thus, residential properties in Japan have been seriously devaluated since the bubble collapsed. At the same time, people's aspirations to home ownership have not declined significantly and remain strong. According to the survey on 'People's Consciousness about Land Issues' conducted by the Ministry of Land, Infrastructure and Transport (2015), the ratio of those who answered that they wanted to own land and a house decreased from 88.1 per cent in 1996 to 79.2 per cent in 2014, but the tendency of most people to desire to own a residential property has not changed significantly. People want to own a home for various reasons. The physical conditions of owner-occupied housing in terms of amenities and floor area are much better than those of rented housing. Home ownership is also regarded as a means of ensuring security in old age. However, those who purchase housing with an expectation and prospect of capital gain have significantly decreased. The survey revealed that the proportion of respondents who agreed that 'land was a more secure asset than saving or stock' dropped significantly from 61.9 per cent in 1994 to 49.2 per cent in 1997 and to 33.2 per cent in 2002 (Ministry of Land, Infrastructure and Transport, 2015). Home ownership in post-bubble Japan is now completely different from before in the sense that most owner-occupied houses have generated capital losses for more than two decades. Nevertheless, many people still aspire to owning their homes. This fact raises a question concerning how we should define the nature of home ownership in the post-growth society.

Housing and generational inequality

Since the beginning of the post-growth era, housing has played an increasingly important role in enlarging and reshaping social inequalities both between and within generations. The impact of radical social shifts on people's fortunes is not equal across generations but concentrates on the younger generation. There has been an obvious disruption in the paths that successive generations follow to acquire property ownership. Owner-occupied housing in Japan, as well as in many other countries, has provided people with the promise of entering mainstream society. Nevertheless, it has become more difficult for younger generations to achieve their aspirations in this regard.

A noticeable phenomenon related to home ownership has been the tendency of younger cohorts to delay entering the owner-occupied housing market (Hirayama, 2012). The level of home ownership for households with a head aged 30 to 34 significantly dropped from 45.7 per cent in 1983 to 31.7 per cent in 1993. This drop was due to inflation in house prices during the bubble period. Even after the bubble burst, the rate of home ownership for the same age group did not recover but decreased further to 28.8 per cent in 2013. As has been discussed, despite the decrease in housing prices, the economic conditions of purchasing housing have become exacerbated with the deflation of real incomes. The casualization of the labor market has led to a decline in regular employment and an increase in non-regular, low-wage employment. This trend has been particularly noticeable for younger generations (Genda, 2001), resulting in discouraging more young households from entering the property market. Furthermore, an increasing number of younger people have delayed marriage, reducing the level of home ownership among younger groups. Despite the drop in the level of owner occupation of younger households, the average home ownership of all households hovered around 60 per cent. This average exists because of an increase in the proportion of the older population having higher home ownership levels. Although owner-occupation has continued to be the dominant housing tenure, routes to the owner-occupied housing sector have become limited for younger people. Generational changes in property ownership opportunities have thus represented shifts in the post-growth society.

Various disparities have become more apparent within younger generations, particularly in terms of marital status and employment status (Hirayama, 2012). There has been a notable increase in young, unmarried adults with non-regular employment who live in their parents' homes indefinitely (Yamada, 1999), while young one-person households have tended to remain in the private rented housing sector, with few prospects of achieving home ownership. Married couples who establish independent households have sought to acquire their own homes. Family households with modest incomes who purchased housing have, however, suffered from the heavy burden of mortgage repayment.

Meanwhile, the housing conditions of elderly people are highly differentiated. In Japan, as in many other societies, owner-occupied housing plays an important role in providing economic security for elderly people, and elderly renters are at a notable disadvantage. Most elderly households in Japan live in owner-occupied housing. On the other hand, those living in private rented housing have suffered from high rents and poor amenities. In addition, landlords may require their tenants to move out. The Japanese government has implemented tenure-biased housing policies aimed at expanding the home ownership sector at the exclusion of the private rental sector, which has resulted in the wretched conditions experienced by many older renters. Moreover, elderly home owners have been differentiated in terms of property ownership. Older people with lower incomes live in owner-occupied houses with lower property values while those with higher incomes tend to own properties with higher values. In addition, higher-income elderly households more often possess a second property that produces rental income (Hirayama, 2010b).

Finally, we should look at the role that intergenerational relations pertaining to housing play in reshaping social inequalities. As Piketty (2014) suggested, as growth slows down, the deployment of family wealth, rather than income flow, becomes more important as a driver of social stratification. For ordinary families, real-estate assets account for the largest portion of family wealth. In post-growth mature economies, there is an agglomeration of a huge housing stock, and how housing wealth is distributed over generations will increasingly dictate how society will be stratified.

Within this context, it is increasingly important to assess what role the inheritance of residential property plays in organizing social inequalities (Hirayama, 2010b; Hoshi, 2001). Owner-occupied housing is not only an asset for owner households but can also be transferred to the next generation. Younger cohorts in Japan have grown up in a more sluggish economy and faced difficulties in acquiring properties by themselves, while a high level of home ownership in the older generation combined with decreasing fertility means increased opportunities for the next generation to inherit residential properties. In the mature post-growth society, therefore, the housing and asset conditions of young households will increasingly be determined by whether their parents own a house and whether they can inherit it. This trend implies that inequalities related to housing properties within the older generation are increasingly being passed down to the next generation (Hirayama, 2010b). It is not likely that housing wealth will be distributed evenly. There will be more families accumulating property assets from one generation to the next, whereas low-income renters whose parents are also renters will continue to be excluded from mainstream society.

Moreover, within the framework of intergenerational relations, parents' assistance related to housing, in addition to inheritance, has played an important role in determining housing opportunities of younger cohorts. This fact means that young people's economic independence has tended to decline in the post-growth era. Together with prolonged stagnation and declining real incomes, the availability of family resources has become increasingly significant in differentiating young cohorts in terms of securing housing. Not all young, unmarried, low-income individuals can live in their parents' homes. Married couples who receive financial support from their parents may find the path to home ownership to be relatively smooth, whereas those who cannot get such assistance must bear the full financial burden of purchasing a house by themselves. In response to enduring recession, the government has substantially reduced taxes on gifts in cases where parents financially support their children in purchasing housing. This measure is expected to simulate economic recovery. The policy of promoting inter-vivo asset transfers has, however, exacerbated inequalities with regard to home ownership opportunities for the younger generation. Wealthy parents are more likely to support their children in entering the home ownership market and obtain tax benefits while parents with limited assets cannot help their offspring to purchase their homes.

Conclusion

Japan's housing system, which developed during the post-war high-speed growth period, was orientated towards mass housing construction and middle-class home ownership. The system was considered to be an effective means of facilitating economic expansion and maintaining social cohesion. Upon entering the post-growth age, however, Japan has experienced demographic, economic and policy changes that have increasingly eroded the foundations of the traditional housing system. In this chapter, we have highlighted the importance of the post-growth context in looking at contemporary housing and social processes. Housing conditions in post-growth Japan have been characterized by the shrinking of housing and mortgage markets, decline in housing affordability due to income deflation and substantial devaluation of residential properties.

In addition, housing has played a growing role in expanding social inequalities both between and within generations. There are now obvious disparities between older cohorts and younger cohorts with regard to housing and home ownership opportunities. Within younger cohort groups, those in precarious employment circumstances have found themselves excluded from the housing and mortgage markets. As for older cohort groups, economic gaps between the home-owning and the non-owning and between the equity rich and the equity poor are becoming apparent. Furthermore, we have pointed out the significance of intergenerational relations pertaining to residential properties in the new dynamics of social stratification. As inheritance and inter-vivo asset transfers increase, inequalities related to housing will be passed on to the next generation.

Housing studies developed during the post-war growth age. This point suggests that housing theories are tacitly premised on 'upward' social changes such as increasing population, rapid economic expansion, housing construction driven by high housing demand and widening access to home ownership. The coming of the post-growth age, however, has brought with it novel social contexts that increasingly require us to re-examine housing theories. For example, it has become necessary to focus on income deflation rather than house price inflation when analyzing housing affordability issues. Within the context of accumulating housing assets in mature economies, we should place more importance on intergenerational relations when exploring changes in social inequalities related to residential properties.

Japan is at the forefront of countries making transitions into post-growth societies. Various other countries in East Asia are expected to follow the housing-related trajectories that Japan has undergone in terms of aging populations and increasingly volatile economies. Singapore, South Korea and China are expected to become 'super-aged societies' with those aged 65 or over making up more than 21 per cent of the population by 2040. The total population of South Korea is expected to begin decreasing in the mid-2020s, and that of Singapore in the mid-2030s. Growth rates are still high in East Asia, but high-speed economic development cannot continue forever, and the region's countries are beginning to experience more volatile economic conditions. Within this context, Japan provides a valuable case study with regard to housing in the post-growth era, and an examination of this country's experience raises questions as to how we should cope with new social realities in reshaping housing systems.

Note

1 1 JPY = 0.0088 USD (as of November 2017).

References

Aalbers, M. (2016). *The financialization of housing: A political economy approach*. London: Routledge.

Crouch, C. (2011). *The strange non-death of neoliberalism*. Cambridge: Polity Press.

Forrest, R. & Hirayama, Y. (2009). The uneven impact of neo-liberalism on housing opportunities. *International Journal of Urban and Regional Research*, 33(4), 998–1013.

Forrest, R. & Lee, J. (Eds.) (2003). *Housing and social change: East-West perspectives*. London: Routledge.

Fujita, Y. & Shionoya, Y. (Eds.) (1997). *Kigyo nai Fukushi to Shakai Hosho [Employee benefits and social security]*. Tokyo: Tokyo University Press.

Genda, Y. (2001). *Shigoto no Naka no Aimai na Fuan [the vague uneasiness of work]*. Tokyo: Chuo Korou Shinsha.

Hirayama, Y. (2007). Reshaping the housing system: Home ownership as a catalyst for social transformation. In Y. Hirayama & R. Ronald (Eds.), *Housing and social transition in Japan*. London: Routledge, pp. 15–46.

Hirayama, Y. (2010a). Housing pathway divergence in Japan's insecure economy. *Housing Studies*, 25(6), 777–797.

Hirayama, Y. (2010b). The role of home ownership in Japan's aged society. *Journal of Housing and the Built Environment*, 25(2), 175–191.

Hirayama, Y. (2011). Towards a post-homeowner society? Homeownership and economic insecurity in Japan. In R. Forrest & N. Yip (Eds.) *Housing markets and the global financial crisis: The uneven impact on households*. Cheltenham: Edward Elgar, pp. 196–213.

Hirayama, Y. (2012). The shifting housing opportunities of younger people in Japan's home-owning society. In R. Ronald & M. Elsinga (Eds.) *Beyond home ownership: Housing, welfare and society*. London: Routledge, pp. 173–193.

Hirayama, Y. (2014). Public housing and neoliberal policy in Japan. In J. Chen, M. Stephens & Y. Man (Eds.) *The future of public housing: Ongoing trends in the East and the West*. New York: Springer Heidelberg, pp. 143–161.

Hirayama, Y. (2017). Selling the Tokyo sky: Urban regeneration and luxury housing. In R. Forrest, S. Y. Koh & B. Wissink (Eds.) *Cities and the super-rich: Real estate, elite practices, and urban political economies*. New York: Palgrave Macmillan, pp. 189–208.

Hirayama, Y. & Izuhara, M. (2018). *Housing in post-growth society: Japan on the edge of social transition*. New York: Routledge.

Hoshi, A. (2001). Gendai kazoku niokeru shisan keisei no kitei yoin [Asset accumulation and contemporary families]. *Jinko Mondai Kenkyu [Journal of Population Problems]*, 57(2), 36–48.

Iwata, M. (2009). Jukyo mondai no tayo na hirogari to homuresu monndai no kozu [Diversified housing losses and the framework of homeless issues]. *Kikan Shakaihosho Kenkyu [Quarterly of Social Security Research]*, 45(2), 94–106.

Ministry of Land, Infrastructure and Transport. (2015). *Tochi Hakusho [White Paper of Land]*. Tokyo: Ministry of Land, Infrastructure and Transport.

Miyake, J. (1996). Jinko kazoku no henka to jutaku jukyu [Demographic and family changes and the housing market]. In S. Kishimoto & A. Suzuki (Eds.) *Kazoku to Jukyo [Family and housing]*. Tokyo: Tokyo University Press, pp. 205–235.

Oizumi, E. (1994). Property finance in Japan: Expansion and collapse of the bubble economy. *Environment and Planning A*, 26(2), 199–213.
Oizumi, E. (2007). Transformations in Housing Construction and Finance. In Y. Hirayama & R. Ronald (Eds.) *Housing and Social Transition in Japan*. London: Routledge, pp. 47–72.
Piketty, T. (2014). *Capital in the twenty-first century*. Cambridge: The Belknap Press of Harvard University Press.
Saito, A. (2012). State-space relations in transition: Urban and regional policy in Japan. In B. Park, R. Hill & A. Saito (Eds.) *Locating neoliberalism in East Asia: Neoliberalizing spaces in developmental states*. Oxford: Wiley-Blackwell, pp. 59–85.
Shinkawa, T. (1993). *Nihongata Fukushi no Seiji-Keizaigaku [Political economy of Japanese-style social welfare]*. Tokyo: San-ichi Shobo.
Streeck, W. (2014). *Buying time: The delayed crisis of democratic capitalism*. London: Verso.
Waley, P. (2015). Pencilling Tokyo into the map of neoliberal urbanism. *Cities*, 32, 43–50.
Yamada, M. (1999). *Parasaito Shinguru no Jidai [The Time of Parasite of Singles]*. Tokyo: Chikuma Shobo.

5 Housing policy challenges and social development in Korea

Seong-Kyu Ha

Introduction

South Korea (hereafter Korea) has suffered from poor housing conditions and a chronic housing shortage, particularly in urban areas. In spite of various policy measures, the housing problem still remains largely unsolved. Analysis of housing policy inevitably involves a consideration of the role of the state in housing provision and consumption. Governments could have subjected housing to free market forces, but in practice they do not. They have policies that influence markets.

For over 40 years, Korean governments have concluded that housing is not something that can be supplied purely according to commercial criteria. State intervention in housing markets has been the norm. Governmental intervention that aims to stabilize prices and to control for speculative demands in the housing market has been a popular approach taken by the Korean government. The housing shortage caused by Korea's continued economic growth and urbanization has often precipitated housing price increases and speculation. Finding solutions for this problem has always been a crucial national challenge while the country was enjoying rapid economic growth, and so increasing housing supply and price control policies were alternately implemented in response to the cycle of the housing market. The housing market and housing policy in Korea are the inevitable products of economic development, political process within the system, housing norms and preferences of the country.

Despite the vast improvement in housing supply and government intervention in housing markets, housing remains a persistent issue in Korea. This apparent paradox arises for several reasons. How have governments played their role in the housing sector? And more specifically, how much did state intervention contribute to social development? We need to analyze whether the policy objectives have been achieved, particularly those related to housing wellbeing and social development.

Housing and social development

The success of a society is linked to the wellbeing of each and every citizen. Social development requires the removal of barriers so that all citizens can journey toward their dreams with confidence and dignity, so they can reach their full potential. With respect to the housing sector in line with social development, to reduce housing poverty, we need to take a social development approach and invest in our people. By investing in people we can reduce housing poverty.

There is no international consensus on measuring poverty. Generally, absolute poverty thresholds are determined by measuring survival needs like food and shelter for different-sized households. The UN's definition of poverty addresses basic needs but recognizes that poverty is multidimensional, with housing poverty[1] as a separate category that can affect – and be affected by – other aspects of a family's life.

In many regions of the world, the number of low-income households far exceeds the affordable housing units available. In Korea, 1,030,000 households (5.4 per cent of total households) do not meet the housing minimum standard in 2016. Housing poverty also can include things like energy and fuel poverty and lack of access to water and sanitation. According to the United Nations Economic Commission for Europe, for example, every tenth person in the European Union lives in a household that was unable to pay utility bills in 2010. Insecure tenure, or the threat of eviction, often lies at the heart of poverty housing, depriving people living in poor circumstances of even the most basic physical, economic and psychological security of adequate shelter. More than 20 per cent of the world's population struggles, on a daily basis, to stay in houses or on land where they live, and more than 80 per cent of the world's population does not have legal documentation of their property rights (Habitat for Humanity, 2016).

It is necessary to clarify the relationship between housing and social development in this chapter. Social development differs from social work and social philanthropy. Social development is dynamic, involving a process of growth and change. Midgley (1999) argues that social development seeks to integrate social and economic processes and cannot take place without economic development. Therefore, social development is defined "as a process of planned social change designed to promote the well-being of the population as a whole in conjunction with a dynamic process of economic development" (p. 27).

With respect to the relationship between social development and housing, social development is not a natural, spontaneous process but instead requires organized intervention in the housing sector. In the Korean economic development process during the last 50 years, social development has been affected by the expansion of government social services and the adaptation of economic planning. The Five-Year Plans are an economic development project of Korea. The plans were designed to increase wealth within Korea and strengthen political stability. General Park Chung-hee seized political power and decided the country should become self-reliant by utilizing Five-Year Plans.[2] Almost half a century has passed since Korea launched its first five-year economic development plan (1962–1966), which triggered the country's compressed economic transformation.[3] As we reflect on the last 50 years, we can see that both change and continuity have followed through Korea's ensuing consecutive Five-Year Plans until 1996 and subsequent market-based policy formation toward an advanced Korea.

According to the World Bank (2005), its experience has identified three operational principles to guide its approach to social development: inclusion, cohesion and accountability. Social exclusion is linked not only to poverty and lack of access to services, but also to housing, which is associated in turn with the increased spatial segregation. Housing-related problems in Korea are multidimensional, including issues of affordability, access, security, quality, and inclusion. Most notably, problems such as housing affordability and compromised housing quality, of varying kinds, are found across housing tenures. They raise concerns about the adequacy of housing tenure as a means of identifying core problems within the housing system.

In 2000, the Millennium Development Declaration (MDG)[4] underlined the importance of social development, basing the MDGs on "certain fundamental values essential to international relations in the twenty-first century". These include: freedom, equality, solidarity, and tolerance. The 1995 Copenhagen Summit on Social Development[5] pledged to "make the conquest of poverty, the goal of full employment and the fostering of stable, safe and just societies" its overriding objectives.

East Asian nations such as the four Asian Tigers (the so-called 'tiger economies')[6] experienced rapid economic development during the last 40 years. The East Asian countries are characterized by a high level of state intervention in social welfare, yet they have experienced rapid economic growth. Social development policy in Korea has developed rather differently from that in either the Western welfare states or the developing nations of the Global South. Perhaps the most important feature of social development in Korea is that social policies have not been divorced from the larger economy as has been the case elsewhere. Social goals have been met by integrating social development with economic development. Vasoo and Lee (2001) argue, "it [social policy] requires that economic development should result in tangible improvements in social well-being through social development" (p. 281).

Korea in recent years has broadened her emphasis from housing for the poor to include providing housing and rent subsidies for specific groups in society with special needs. These include diverse groups such as the elderly, single people and the physically handicapped. Housing policies for vulnerable social groups are giving more and more priority to low-income communities. The concentration and segregation of disadvantaged groups in poor neighborhoods has been viewed as one of the challenges of social development in line with housing policy in Korea. Intervention in the housing market seems to be a generally accepted solution to break the vicious circle of the reproduction of poor neighborhoods. The key challenge is to develop various subsidies and affordable housing programs to help reintegrate these social groups into society, particularly through equal access to public services. Housing programs for vulnerable social groups are getting embedded in the national housing and social programs.

Housing policy challenges and social development

Government intervention in the housing sector is mainly based on efficiency and equity grounds. From the viewpoint of efficiency, intervention is for reducing market failure in housing markets and enhancing efficient allocation of resources. On the other hand, from the equity viewpoint, intervention is for providing adequate housing services for those who cannot afford housing on their own in housing markets. Housing stability and improvement of housing conditions have been the major goals of housing policies in Korea. For the last several decades, there have been substantial legislative and policy challenges and changes. It is necessary to explore how housing policy has changed over time along with policy challenges and what type of housing programs have been implemented for social development.

Expending housing production

In order to tackle the housing shortage, the government started to promote a new housing supply plan, known as the Two-Million Housing Unit Construction Plan (1988–1992), which proposed to construct considerable quantities of high-rise apartments in five new towns near the capital city of Seoul. The Plan primarily focused on mitigating housing shortages by expanding the housing production on a large scale, which would also reduce housing prices and stabilize the housing market. Compared to the previous housing provision, the plan was unprecedented in Korea. Table 5.1 presents the planned housing supply program of the Two-Million Housing Construction Plan. As a result of the construction plan, the quantity of housing supplied by the private sector was 1,432,000 units, which were much more than the initial plan of 1,100,000 units. In contrast, the supply of public sector housing (711,000 units) was substantially lower than the initial plan (900,000 units). In terms of housing supply, the Two-million Housing Construction Plan was successfully implemented; the government intervened in the housing market by expanding housing production on a massive scale.

The stock of public rental housing accounted for only 3.5 per cent in 1980. Between 1971 and 1980, 64,947 apartments were constructed for public rental housing. However, these were very short-term rental houses: after one or two year's mandatory rental period, they were sold off to the occupants. The public agencies in charge of public housing including the Korea National Housing Corporation (KNHC) had to recover the cost in a short period of time. The housing policy change through the Two-Million Housing Construction Plan was the adoption of a welfare housing system, which was permanent public

Table 5.1 Two-million housing unit construction plan, 1988–1992 (unit: thousand)

Category		*Total*	*Year*			
			1988/1989	*1990*	*1991*	*1992*
Public sector	Permanent rental houses	250	43	60	70	77
	Welfare houses for the laboring class	150	–	40	–	60
	Rental houses for the working class	100	–	20	–	50
	Long-term rental houses	150	91	25	80	14
	Small-sized houses for sale	250	142	55	40	13
	Subtotal	900	276	200	210	214
Private sector	Subsidized houses for sale	601	241	120	120	120
	Houses for sale	499	262	130	70	37
	Subtotal	1,100	503	250	190	157
	Total	2,000	779	450	400	371

Source: Ministry of Construction and Transportation (2002); Korea Research Institute of Human Settlement (2013).

rental housing for the low-income households. A permanent public rental housing program represented the beginning of a social housing tradition directed at low-income families. In addition, the government tried to provide affordable housing for the housing poverty group.

During the early 1980s, the government began to expand public rental housing. But public rental housing with respect to welfare purposes for the housing poverty group, such as Permanent Rental Housing in Korea, was provided in only a small quantity. Permanent Rental Housing is targeted at people in the lowest-income bracket, namely residential protection target groups and self-supporting protection target groups as defined in accordance with the Livelihood Assistance Law. They are households who are unable to make a living, due to wage earners being either too young or too old. Also entitled to the Permanent Rental Housing are households displaced by urban redevelopment projects and relief recipients (normally families of patriots and veterans martyred for the nation). Rent is usually 25–30 per cent of market prices. The 50-year Rental Housing was designed for relief recipients, the urban poor displaced from urban redevelopment projects and disabled persons who do not own homes. However, the government has not built this type as well as Permanent Rental Housing since the early 1990s because of heavy financial burdens on the government. Five-year Rental housing was designed for low-income households who do not own houses and have saved a certain amount of money with the Housing Subscription Savings. By the mid-1990s, 100,000 units were provided annually. Company Employee Housing is built for household heads who do not have their own homes and are employed in companies with 5 or more employees. Industrial companies buy homes at subsidized prices and then rent to their employees. National Rental Housing was newly designed by the Kim Dae-Jung administration in 1998. There were two types: 20-year rental housing for households in the lowest 20 per cent income bracket and 10-year rental housing for those in the lowest 40 per cent. Public rental housing excluding 5-year rentals that cannot be regarded as social housing currently accounts for only 5 per cent of the total housing stock (Table 5.2). This amount is compared to 20 to 30 per cent in some European countries and is often blamed for the unstable rental housing market for low-income households in Korea.

Happy housing projects

The Korean economy has developed faster than any other nation's, but expensive residential costs have been identified as one of the most serious social problems. Having one's own home is becoming a distant dream for young couples and low-income families. The Park Geun-hye administration,[7] which came up with a vision of national happiness, has been focusing its energy on "Happy Housing" projects providing 150,000 units until 2017. The main target of the Happy Housing Project is on young people in their 20s to 30s. Some 60 per cent will be first provided for university students, young workers and newlywed couples; another 20 per cent will be allocated for people in vulnerable groups such as the disabled. Non-home owners will be able to take the remaining 20 per cent. Designed as an eco-friendly complex, the Happy Housing Project will lure social entrepreneurs to expand job opportunities for residents. It was also planned to have individual themes such as the environment, university, communication, and multicultural families.

It will first be implemented in seven districts, which were chosen based on the review of a number of factors including demand, population structure, market conditions and local conditions. Also, the seven designated districts are planned to become places for people to find jobs and receive welfare and cultural services. In order to achieve the goal of the Happy Housing project finding suitable land is the biggest challenge. Due to the high land prices in the urban area, supplying affordable public rental house there is unrealistic. In addition, residents worry that the newly constructed public rental houses will densify the region and enhance the negative image of the community, which is the biggest supply obstacle. To solve the problem, the government tries to utilize state-owned land in the city area to enhance supply efficiently. Therefore, the Korean government selected seven sites of railway stations and their areas in the Seoul metropolitan area, mainly government-owned land where there are strong housing needs. More than 10,000 residential units will be provided to people in need of shelter in the central city areas by developing state-owned land such as railway properties and reservoirs. The combined size of the designated areas is 489,000 square metres (5.3 million square feet). The areas have convenient access to public transportation and are equipped with educational and commercial facilities (Ministry of Land, Infrastructure and Transport, 2017).

The Happy Housing project was an election pledge by President Park to tackle the country's chronic housing problems. The project is similar to the previous government's Bogeumjari Housing Project,[8] but the new program

Table 5.2 Public rental housing stock in 2014

Types (in Korean)	*Permanent Rental (Young-gu)*	*Public rental (Gong Gong)*	*National rental (Kookmin)*	*Company employee rental (Sawon)*	*Public & private Keunseol rental*	*Public purchase and rent (maeip)*	*Total*
Rental period	50 years	50 years	30 years	(50/10/5 years)	5 years	3/5 years	–
No. of housing units (unit: thousand)	193	106	521	30	385	358	1,709

Source: Ministry of Land, Infrastructure and Transport (2015).

provides housing in central urban areas instead of rural areas. Construction will be undertaken by the Korea Land and Housing Corporation and the SH Corporation. The Korean Railroad Corporation and regional offices will provide the central government with their railway and reservoir properties.

However, in the process of gathering public opinion, some experts and local residents voiced objections as they thought constructing on these sites was impractical and a top-down policy as the government did not consider community views. This project was also criticized for not having close cooperation between central and local governments.

In addition, the current government is switching from the integrated allowance system to a customized separate allowance system by revising its National Basic Livelihood Security System (NBLSS) allowance system. The integrated allowance system is a system that supports the poor in terms of their income level without considering their housing situation. The focuses of the new housing are reducing the rent burdens of households, improving housing conditions and efficient use of existing housing stock. The allowance system was expected to increase the number of eligible households by raising the income limit from 730,000 KRW[9] to 970,000 KRW (home-owning households included) and to increase the subsidy amount by raising average monthly subsidies from 80,000 KRW to 110,000 KRW.

From 2010 on, this allowance system was used under the name 'housing voucher', but it provided a subsidy not as a coupon but in cash in Seoul. In fact, it is regarded not as a housing voucher but a 'housing allowance' or 'income subsidy.' It gives aid to households classified in the bottom 20 per cent of income (whose recognized incomes represent at most 150 per cent of the minimum cost of living), excluding the recipients under the National Basic Living Security Act. In order to be successful with this system, it is necessary to ensure adequate standards of housing conditions for beneficiaries and to create an effective housing welfare delivery and monitoring system.

Control of housing price

In the late 1970s, housing prices of new apartments were skyrocketing. Housing price control first applied to public housing. A price ceiling is a government-imposed price control or limit on how high a price is charged for a product. Governments use price ceilings to protect consumers from conditions that make necessary commodities unattainable. The regulation was introduced to prevent the spread of speculation, to lower the parcel price (sale price) and to secure easy buying by non-home owners. However, it began to apply to private housing built in accordance with the Housing Construction Promotion Law (HCPL) in 1977 when rises in new house prices became an issue. Developers were blamed for appropriating excessive profits from the buyers while they were benefiting from various favors including exemption or reduction of the VAT, corporate tax and transfer tax. The sharp rise in house prices also raised public criticism of developers for their excessive profiteering. The Seoul city government announced a plan to control such profiteering through administrative guidance and to consider setting a price ceiling in accordance with the Price Stabilization and Fair Trade Law.

In the first period of price ceiling regulation, a parcel price was set uniformly regardless of its land price. Later, the regulation was relaxed from the uniform price ceiling to taking into consideration the production costs such as land price, construction costs, etc. The government set the level of each component of production costs: standard building cost, reasonable profit and so on.

This price ceiling remained almost unchanged until 1989, when a new system, 'the Cost Linking System', replaced it. Under the new system, new house prices built in accordance with the Housing Construction Promotion Law (HCPL) were to be assessed and approved by the government based on land cost and the standard construction cost. The price control was criticized for encouraging housing speculation. This was because there was always excess demand for houses under price control because the selling prices of new houses were in many cases far below the market prices. The gap between the purchase price and the ceiling price (sale price) became capital gains for purchasers. This gap had brought about the prevalence of housing speculation. The government thus had to regulate the sale of all new houses as well. The HCPL prescribes additional regulations for this new measure, too. New houses built with the financial support of the National Housing Fund and Korea Housing Bank loans are to be sold preferentially to housing subscription depositors who have not been home owners for a certain period of time. Those who have acquired housing with support from the Fund or the Bank are not allowed to sell or rent their new houses before a stipulated period of two years.

However, the price controls on the new apartment submarket produced negative impacts and consequences (e.g., speculation). The price ceiling regulation was thus relaxed gradually, and finally the price control on new apartments was abolished in April 2015. It was mainly due to the global financial crisis and the collapse of housing prices. The government wanted to boost the housing sector in an attempt to stimulate an economic recovery. But the system of price control in public sector housing for sale is still enforced.

Urban and housing renewal/regeneration

Urban and housing regeneration is considered a good way of solving diverse problems in Korea such as the lack of housing and deteriorated living environment. The urban renewal projects were accompanied by the booming housing market of the 1980s, when high-rise apartments became the most popular type of housing in Korea. There is no doubt that the urban and housing renewal projects have contributed to the increase in housing stock. The number of redeveloped housing units was more than twice the number of original units. Also, the average floor area in the newly developed units almost tripled the average area of the old units. It should be noted here that these regeneration projects were led by property owners excluding tenants.

There are several reasons why I raise the issue of urban and housing renewal policy in line with social development. First, the scale of impact of the housing renewal projects is far greater than for other projects. Low-income households have faced difficulties in resettlement due to rising housing prices and housing speculation, which made a wide array of low-income families suffer. Second, the purpose of renewal focused on dealing with physical improvement rather than enhancing low-income families' housing stability. Third, housing and urban renewal projects have caused the most acute outcomes such as speculation, increasing housing prices, eviction, the low rate of resettlement of original residents and broken social ties between residents.

Additionally, the aforementioned projects led to social conflicts between rich and poor residents in the same communities. After the projects had been enforced, different levels of income groups moved into the same neighborhood and changed the way that the original residents used to live before. Conflicts between home owners who live in private housing and tenants who lived in public rental housing are quite often seen in the newly redeveloped areas. According to a recent survey, more than half of the respondents did not want to live with those who belonged to different income strata, cultures and ethnic minorities (Ha, 2008).

Furthermore, if some renewal neighborhoods had public rental housing, the rent was extortionate for low-income people, especially the original residents who used to live in the area before the project was implemented. Thus, low-income tenants had no choice but to move out of their neighborhoods. Survey evidence suggested that nearly 80 per cent of the original residents were displaced in the process of regeneration projects. A considerable sum of the development profits from the projects was passed to outside investors (speculators) and real estate brokers instead of going to the original residents in redevelopment sites (Seoul City's Advisory Committee for the Improvement of Residential Environments, 2009; Shin, 2008). Urban regeneration has been closely linked with the issue of housing reform for the urban poor since the early 1980s. The urban regeneration policies could have been successful if many of original residents had remained in the redeveloped area after the completion of the project. But in reality, the rehousing in-situ ratio for original residents was very low, and most of them had to move to areas where housing costs were lower.

The government was too late to change and amend existing laws and policies to prevent the aforementioned problems. Fundamentally, this continuing vicious circle, that is, social problems occurring as a result of the property-led regeneration, is difficult to break without examining housing regeneration and low-income housing policies. Thus, housing and urban renewal programs continue to pose obstacles to social development in Korean society.

In 2002, the so-called New Town Program was also initiated, which is a much more aggressive housing redevelopment and urban renewal program than the previous Joint Redevelopment Program. The New Town Program focused on comprehensive development with a large-scale master plan. In the midst of the post-financial crisis housing market boom, this program was welcomed by many property owners of low-income residential areas, especially in Gangbuk (northern part of Seoul). In total, 26 areas covering 23.8 square kilometers including about 850,000 people (about 8 per cent of the total population of Seoul) were designated in Seoul. Among a total of 350,000 households, 69 per cent (240,000 households) were counted as tenants (Jang & Yang, 2008).

The city government has become an active enabler of the New Town Program, requesting the enactment of a new law to provide legal bases for the deregulation of the Urban Planning Law and the Building Code to attract the private sector and special subsidies from the central government. The Special Promotion Act of Urban Renewal was legislated for the program in 2006. This Act overrides the authorities of the existing laws related to urban planning and development. It is notable that the Special Promotion Act encourages large-scale housing and urban renewal. The criteria for the project area designation are not as tight as those for the Joint Redevelopment Program. Areas not previously considered deprived areas can now be thus designated. The Seoul city government has designated 241 renewal promotion districts in its 26 New Town Project areas (Kang, 2012).

In February 2014, Park Won-soon, Mayor of Seoul, presented the vision of the urban and housing renewal policy. The Seoul Metropolitan City Government has unveiled a new plan for urban living space improvement. The city government will also establish a new agency to be in charge of the job of urban renewal at the cost of 1 trillion won for the next four years. The main feature of the city government's new plan is to move its focus away from bulldozing an area for cookie-cutter apartment buildings and toward revitalizing village communities, local job creation, and local identity preservation. In addition, future urban improvement projects will be undertaken with local residents' requirements and demands fully incorporated. The 1-trillion-won budget for the new agency will also be used to build and restore roads and local parks, common-use facilities, old city walls, and other historic sites located in residential areas. Mayor Park said, "from now on, preserving local identity so that people in communities can work, enjoy, and raise children together is the main vision of our urban renewal initiative for the city" (*The Korea Economic Daily*, 2014). Having discussed the housing policy changes and spatial policy related to housing development, let us turn to the social ramifications.

Social exclusion

There are some crucial and divisive issues in housing and social development. The most important emerging issue is "social exclusion". The idea of "exclusion" is not new, but it has only recently become important in Korea in terms of social development. In this study "exclusion" is used to refer to the process that keeps people from being fully integrated into society and not supported by it. This concept looks at the relationship between housing and social exclusion. Housing is an important element in exclusion, and housing policy has a major contribution to make to strategies concerned with social inclusion. Social exclusion manifests not only in levels of income, but also in matters such as health, education, access to services, housing and debt (Power & Wilson, 2000; Spicker, 1998)

One of the most striking cases of exclusion in Korea is related to people who are low-income tenants and slum dwellers. The governments do not take care of them quite as comprehensively as we might like to think. Housing is itself a cause of exclusion; the places where people live can be part of the reason for their social rejection. Poor housing is a form of deprivation; it may also be an indicator of other problems. People are not necessarily poor because they live in poor housing; but if they are poor, they are more likely to have to live in poor housing. Housing is largely distributed according to economic criteria, which means that people with least command over resources are also disadvantaged in their ability to obtain housing. The distribution of public housing is an important corrective, but the proportion of public housing stock is much less than the number of poor families in Korea. People with few resources are probably not going to be able to obtain or afford decent housing.

Since the early 1980s, the Korean government has developed public rental housing programs with the goal of contributing to social integration through providing the housing poverty groups with decent and affordable accommodations. However, there has been a growing concern that public rental housing estates have become stigmatized and isolated from the outside at a local level. The phenomenon of conflict between public rental housing estates and local people not living in public rental housing estates has been debated under the term "social exclusion" not only by Korean academia but also by the government (Kang, 2015). Public housing communities are, in general, described with negative social images. From the perspective of the "neighborhood effect of poverty concentration", disadvantaged communities are usually characterized as having dominant deviant norms, limited social networks and weak neighborhood attachment. Some local governments in Korea opposed to public rental housing mainly due to its negative images.

The author's survey findings demonstrate that there is a growing stigma against the poor and social exclusion (Ha, 2008). In this survey, a systematic random sampling was used to choose the households to be interviewed, and 300 households (150 for social housing and 150 for non-social housing) were interviewed in the three study areas (Beondong, Jungae and Deungchon) in Seoul. This kind of social bias is likely to escalate the neighborhood opposition to construction of social housing estates. Public housing was thought to have a negative impact on the neighborhood. One of most crucial findings is that the mixture of social housing and non-social housing within the same community has become an issue. More than half of non-social housing residents have a negative view of the social mix. They think that it is desirable for similar income groups to live in the same neighborhood. The negative view means "birds of same feather flock together".

Similarly, about 44 per cent of social housing residents agreed that it is undesirable to mix social housing and nonsocial housing within the same estate.

Some occupants (25.4 per cent) of social housing responded that neighboring people did not have friendly feelings toward social housing. It is noteworthy that 28.6 per cent of people in social housing have feelings such as "sometime feeling a shame to live in social housing." It is also important to point out that children who live in social housing were often discriminated against by their peer group at school (16.3 per cent). Since the early 2000s, several stories of students (particularly social housing residents) being maliciously alienated by their fellow classmates in school have shocked the nation (Ha, 2008).

The *Hankeoyrae 21* reported a case in which barbed wire was set up to close a road between public rental apartments and privately owned apartments within a housing estate in Seoul (Ryu, 2006). The public housing residents requested to use the road. The owners of the private housing, however, disliked having public housing residents pass through their complexes, including children in public housing who might take the road to school to save ten minutes. The residents of public rental apartments were considered a nuisance to the middle-income neighborhood. This resulted in a lawsuit being filed by residents of the public rental apartments. It is important to examine why these kinds of social problems arise. Applying the concept of social sustainability to low-income communities in urban Korea would require mobilizing residents and their governments to strengthen all forms of community capital. The government has to be seen to be allocating housing fairly and equitably. The housing authority has to balance a number of competing objectives. The Korean government must undertake to get the best use of the available stock and to meet social needs and preferences while avoiding geographic polarization by age, family structure, and income.

Minimum housing standard

According to the census results, the housing shortage has decreased considerably due to government policies to promote housing construction.[10] The results of the Population and Housing Census have been used to measure the quality of dwelling conditions of households as well as the quantity of housing in Korea. The Korean Government has concentrated its effort on housing supply rather than the improvement of housing condition for several decades. As a result, the housing supply ratio (housing stock/households in demand of housing) has gradually increased from 71.7 per cent in 1985 to 103.5 per cent in 2014. To further enhance housing development, the policy makers of Korea have moved their focus from providing more housing to improving housing conditions.

The objective of the housing minimum standards is to measure the quality of households and support the improvement of housing conditions of the households under the housing minimum standards (Gann, Wang & Hawkins, 1998; Ha, 2002). For the improvement of housing conditions and wellbeing, the Korean Government introduced housing minimum standards in 2000. The housing minimum standards also embody Article 35 of the Constitution of the Republic of Korea that enacts the citizen's right to a healthy and agreeable environment and give responsibility to the state to ensure comfortable housing for all of its citizens.

The minimum standards of housing in Korea consist of four components: a bedroom standard, housing facilities standard, floor space standard and environmental standard. The total number of households under the housing minimum standards is 1,030,000 in 2016, which accounts for 5.4 per cent of total households. According to an analysis by Jeon & Lee (2007), households under the minimum housing standards possess different characteristics according to region (urban/rural), household head's age, type of occupancy, type of household, and year of housing construction. Housing conditions of residents residing in rental houses are worse than those of house owners. The dwelling conditions of renters are weak in respect to the number of bedrooms and floor space. According to household types, most family households have enjoyed good housing quality. Only 9.5 per cent of family households are under the minimum housing standards. Among one-person households and households with no blood-ties, 21.2 per cent and 22.0 per cent of the households are under the housing minimum standards, respectively. If a household head is female or the construction is old, the household has more of a probability of being under the minimum housing standards.

The minimum housing standards were specified in terms of the number of rooms and floor area, differentiated by the size and composition of households. The minimum standards were upgraded in 2011 by increasing the minimum floor area as well as requiring a modern kitchen, toilet and bath/shower. Even though the housing minimum standards are important criteria to judge the citizens' housing condition, there are no specific regulations on improving the poor housing conditions. In order to achieve balanced and greater social development in Korean society, the government must first establish a medium- and long-term policy for households that do not meet the minimum housing standards. Given the financial limitations, priority should be given to vulnerable groups such as those with a disability who are below the minimum standards.

Homelessness

There are several types of poor and illegal accommodations for the poor people in urban areas (Table 5.3). Goshiwons are tiny one-room closets with no windows but space for only a bed and a desk. These are very cost-effective, especially for students and low-income single-person households. Jjogbang, meaning very small and narrow rooms, averages approximately 3.3 square metres. Most Jjogbang suffer from a lack of facilities and poor services, and their tenants are mostly poor and homeless people. Vinyl houses are constructed of layers of thin wooden boards with vinyl covering on the outside. Most vinyl house occupants are poor tenants who have been forcibly evicted from housing renewal areas, particularly during the 1980s and 1990s. Most simply settled on vacant hillside areas or public open spaces, without any rights to land ownership or building permits.

Table 5.3 Number of population of vulnerable social groups

Type	*Street homeless*	*Vagabond facilities*	*Shelters for the homeless*	*Emergency shelters*	*Multiplex premises*[1]	*Jjokbang*	*Inns*	*Gosiwon*	*Vinyl house areas*	*Vinyl houses, containers, mud huts, etc.*	*Total*
Total	2,689	8,160	2,636	508	62,453	6,214	15,440	123,971	6,914	32,053	261,038

Source: Ministry of Health and Welfare (2012).

Note

1 On the multiplex premises is the *jimjilbang*, which is a large, gender-segregated public bathhouse, furnished with hot tubs, showers, Korean traditional kiln saunas and massage tables. Most *jimjilbangs* are open 24 hours, and one can sleep overnight and enjoy the bathhouse and sauna.

Homelessness is the condition of people without a permanent dwelling. People who are homeless are most often unable to acquire and maintain regular, safe, secure and adequate housing. As shown in Table 5.3, the number of this vulnerable social group was more than 260,000 in 2011. However, the street homeless (2,689 persons) are a rather small number compared to other types of housing poverty groups. By region, Seoul has the largest number of homeless people with 3,304. The capital saw a whopping 67.1 per cent rise from 2,671 homeless people to 3,304 over a two-year period (2009–2011). The rapid rise indicates a jump in the number of people experiencing economic difficulties due to the recession (particularly after the outbreak of the world financial crisis in 2008), which prompted a wider polarization of wealth, unemployment and a stagnant property market. In reality, it is difficult to obtain accurate data on homelessness. It is in fact necessary to conduct a comprehensive nationwide investigation to reveal the real situation, and the government should act accordingly.

As a matter of fact, in order to tackle housing poverty, the government has been trying to provide accommodation for the vulnerable social groups. In Korea, the government enacted the Law on Welfare of the Homeless and Supporting Self-sufficiency in 2012. Its Article 2, Section 1 stipulates that "homeless people" includes: 1) people who for a considerable period of time have no housing, 2) people who live at a facility for the homeless for a considerable period of time and 3) people who live in undesirably poor housing for a considerable period of time.

According to the Seoul Metropolitan Government (2014), about five per cent of the homeless were women. The most common place to sleep was underground spaces, followed by parks, in buildings and common areas in the street. The city government is trying to persuade homeless people to sleep in shelters, but last year only 3,605 people signed up for the shelters. The city government gives homeless people who live in shelters housing assistance and runs self-support programs and vocational training.

Conclusion

An analysis of housing policy inevitably involves a consideration of the role of the state in housing provision and consumption. Korean governmental intervention that aims to stabilize prices and to control for speculative demands in the housing market has been a popular approach. The direct and indirect participation of the Korean government in housing has over the years considerably increased through enacting an impressive number of regulations, laws and plans for the construction of dwellings, land development and other activities. Institutional foundations to make policies for the housing-underprivileged were prepared. The Act on Support for Underprivileged Group, Disables Persons and Age was enacted in 2012. This act provided the basis to prepare policies such as the planning and monitoring for the housing-underprivileged. Furthermore, the Act on Support for Welfare and Self-Reliance of the Homeless was enacted in

2011 to provide housing programs and support polices for the homeless (MOHW, 2016). However, the policies for the housing-underprivileged have not been systematized and formulated yet.

It should go without saying that housing is essential to wellbeing, even life. It is so much more than a physical space or structure. Housing is where we develop our first social relationships, it ties us to our communities and it is connected to our livelihoods. Just as housing goes beyond four walls and a roof, housing poverty is not about just the lack of a house. In order to broaden the response to vulnerable social group including homelessness and thus effectively address it, we need a paradigm shift. We have to move away from an exclusive focus on the individual circumstances leading to and arising out of vulnerable social groups toward a focus that recognizes the structural causes of the groups as well as its individual dimensions. Unfortunately, the crisis of vulnerable social groups and the recognition of housing as a human right have not been key issues in Korea, leaving a wide gap between housing policy and government obligations. Without doubt, building houses will be part of any strategy to eliminate housing poverty. But only a human rights response can address what is fundamentally a human rights failure.

The basic aim of the government's housing policy is to ensure that every family can obtain a dwelling of good standards located in an acceptable environment, at an affordable price or rent (National Law Information Center, 2016). Despite the vast improvement in housing supply and government intervention in housing markets, housing remains a persistent and divisive issue in Korea. We would conclude that the goal of housing policy as stated in Housing Act has not been fairly achieved. In terms of social development, the government should recognize that housing policies should surely address not only the provision of bricks and mortar. The state intervention should contribute to social development particularly for the vulnerable groups of people. Overcoming housing poverty is not just a matter of getting economic and housing policies right; it is also about promoting social development, which empowers people by creating more inclusive, cohesive, resilient, and accountable institutions and societies.

Notes

1 Housing poverty refers to conditions that do not meet the minimum housing requirements. Minimum housing requirements are listed as "tenure security, affordability, adequacy, accessibility, proximity to services, availability of infrastructure, and cultural adequacy."

2 Park Chung-hee became President of South Korea after a military coup (1961) and was assassinated in 1979.

3 The first plan sought to expand electrical/coal energy industry, emphasizing importance on the infrastructure for establishing a solid foundation, agricultural productivity, export, neutralize balance of payments, and promote technological advancements. Korean economy observed a 7.8 per cent growth, exceeding expectations, while GNP per capita grew from 83 to 125 USD (1962–1966).

4 On September 8, 2000, following a three-day Millennium Summit of world leaders at the headquarters of the United Nations, the General Assembly adopted the Millennium Declaration. A follow-up outcome of the resolution was passed by the General Assembly on December 14, 2000 to guide its implementation. Progress on implementation of the Declaration was reviewed at the 2005 World Summit of leaders.
5 At the conclusion of the World Summit for Social Development, held March 6–12, 1995 in Copenhagen, Denmark, Governments adopted a Declaration and Program of Action, which represents a new consensus on the need to put people at the center of development.
6 The Four Asian Tigers, or Little Dragons, are the economies of Hong Kong, Singapore, South Korea, and Taiwan, which underwent rapid industrialization and maintained exceptionally high growth rates (in excess of 7 percent a year) between the early 1960s (mid-1950s for Hong Kong) and 1990s.
7 Park Geun-hye was the President of Korea from 2013 to 2017. Park was the first woman to be elected as President in Korea and served the eighteenth presidential term.
8 *Bogeumjari* housing project refers to construction of public housing for the purpose of rental or sale at affordable price to support low-income households. In May 2010, The *Bogeumjari* Housing Project has been confirmed to supply 140,000 (78 per cent) and 40,000 (22 per cent) houses to the metropolitan and non-metropolitan areas respectively after designating new districts, developing existing residential areas, and considering floor area ratios for redevelopment and reconstruction (See Korea Land and Housing Corporation [2017] and Seoul Information Communication Plaza [2017]).
9 1 KRW = 0.0009 USD (as of November 2017).
10 The Population and Housing Census of Korea has been carried out on a quinquennial basis since 1925. The census provides a "base population" for the population projection for the whole country as well as for the provinces.

References

Gann, D. M., Wang, Y. & Hawkins, R. (1998). Do regulations encourage innovation? The case of energy efficiency in housing. *Building Research & Information*, *26*(5), 280–296.

Ha, S. K. (2002). The urban poor, rental accommodations, and housing policy in Korea. *Cities*, *19*(3), 195–203.

Ha, S. K. (2008). Social housing estates and sustainable community development in South Korea. *Habitat International*, *32*, 349–363.

Habitat for Humanity. (2016). *7 things you should know about poverty and housing*. Available at www.habitat.org/magazine/article/7-things-you-should-know-about-poverty-and-housing, accessed on November 5, 2016.

Jang, N. J. & Yang, J. S. (2008). *Key issues and improvements of new town project Seoul*. Seoul: Seoul Development Institute (In Korean).

Jeon, S. A. & Lee, J. W. (2007). Households under the minimum housing standard. *Paper presented at 23rd Population Census Conference*, April 16–18, Christchurch, New Zealand. Available at http://unstats.un.org/unsd/censuskb20/Attachments/2007KOR_CensConf-GUID845b2545091b406db71fe56f6fd0f52e.pdf, accessed on September 16, 2016.

Kang, T. S. (2015). *The existence and causes of social exclusion on public rental housing estates in South Korea: The universalism of the undeserving poor*. Doctoral thesis submitted to University of Birmingham.

Kang, W. (2012). New town project of Seoul, Korea: an evaluation and future direction. *Journal of the Korean Urban Management Association*. *25*(4), 153–173.
Korea Land and Housing Corporation (2017). *LH official website*. Available at www.lh.or.kr, accessed on October 6, 2017.
Korea Research Institute of Human Settlement (2013). *Housing policy*. Anyang: KRIHS.
Midgley, J. (1999). *Social development, the development perspective in social welfare*. London: Sage.
Ministry of Construction and Transportation. (2002). *Housing white paper: Past and present of housing stability*. Seoul: MOCT.
Ministry of Health and Welfare (MOHW). (2012). *A report on national actual condition survey on vulnerable social group 2011*. Seoul: Ministry of Health and Welfare.
Ministry of Health and Welfare (MOHW). (2016). *Comprehensive plan for welfare and independence support for the homeless*. Seoul: Ministry of Health and Welfare.
Ministry of Land, Infrastructure and Transport. (2015, October 20). *Rental housing statistics*. Available at http://stat.molit.go.kr/portal/cate/statView.do?hRsId=37&hFormId=1248&hDivEng=&month_yn= /, accessed on April 16, 2016.
Ministry of Land, Infrastructure and Transport. (2017). *Introduction to happy housing*. Available at www.molit.go.kr/happyhouse/info.jsp, accessed on October 23, 2017.
National Law Information Center. (2016, April 10). *Housing act*. Available at www.law.go.kr/eng/engMain.do /, accessed on October 21, 2017.
Power, A. & Wilson, W. J. (2000). *Social exclusion and the future of cities*. Centre for Analysis of Social Exclusion, London: London School of Economics.
Seoul City's Advisory Committee for the Improvement of Residential Environments. (2009). *Unpublished document for a public hearing*, January 2009. Seoul: Seoul Metropolitan Government.
Seoul Information Communication Plaza. (2017). *Bogeumjari housing*. Available at http://opengov.seoul.go.kr/civilappeal/2895919, accessed on October 6, 2017.
Seoul Metropolitan Government. (2014, June 14). *Center for homelessness*. Available at www.homelesskr.org/, accessed on September 15, 2016.
Shin, H. B. (2008). Living on the edge: financing post-displacement housing in urban redevelopment projects in Seoul. *Environment and urbanization*, *20*(2), 411–426.
Spicker, P. (1998). *Housing and social exclusion. A discussion paper*, Dundee, Scotland: University of Dundee.
Ryu, W. J. (2006, March 29). Demolish the wall between the better-off and the worse-off, No. 603. *The Hankyoreh, 21*. Available at http://legacy.h21.hani.co.kr/section-021003000/2006/03/021003000200603290603084.html, accessed on April 5, 2006.
The Korea Economic Daily. (2014, February 26). Seoul city gov't unveils new urban renewal vision. Available at http://english.hankyung.com/news/apps/news.view?c1=01&nkey=201402261345191, accessed on October 30, 2016.
World Bank. (2005). *Empowering people by transforming institutions: Social development in world bank operations*. Washington, DC: World Bank.
Vasoo, S. & Lee, J. (2001). Singapore: Social development, housing and the Central Provident Fund. *International Journal of Social Welfare*, *10*, 276–283.

6 Changes in housing policy, housing wellbeing and housing justice in Taiwan

Chin-Oh Chang and Bor-Ming Hsieh

Introduction

Taiwan's housing market has developed some unique characteristics in the Asian context. Over the past several decades, the government has made many efforts to promote home ownership; therefore, Taiwan has achieved an amazingly high home ownership rate, reaching 85.4 per cent of total housing stock in 2016 (Directorate-General of Budget, Accounting and Statistics, 2017). However, problems hide behind the high home ownership rate. Since the government spent a vast majority of its housing resources on the owner-occupied housing market, the rented housing markets in both the public and the private sectors are relatively small and vulnerable. Private rented housing units account for less than 15 per cent of the total housing stock. For a long time, public housing policy aimed to promote home ownership among middle and low-income households; as a result, most public housing units were for sale. Public rented housing (or so-called social rented housing) units accounted for only 0.09 per cent of housing in 2016.

Moreover, Taiwan's housing market has experienced great fluctuations in the past decade. After the breakout of Severe Acute Respiratory Syndrome (SARS) in 2003, the housing price increased to more than double in metropolitan areas, which has caused a serious affordability problem for many middle and low-income households. However, with high housing prices, the vacancy rate was also high, reaching 19.3 per cent (1.65 million units) in 2010 (Directorate-General of Budget, Accounting and Statistics, 2012). The three high conditions in Taiwan's housing market–high home ownership rate, high housing prices and high vacancy rate–reveal the government's long-term skewed housing policy and ineffective management and intervention in the housing market. A unique type of small government and big market has been developed due to the government's long-term failure to build a well-functioned mechanism in the housing market. Taiwan has experienced acute speculation in the real estate market, and some people have made fortunes from it.

To cope with high housing prices and to restrain speculation in the housing market, the governments led by the Kuomintang or Democratic Progressive Party have implemented a series of financial strategies and credit controls in the housing market as the government did successfully in the late 1990s. However, these financial strategies have had limited effects on the housing market. The community has increasingly realized that the problem in the housing market results from unsound decisions in housing policy development and housing taxation. The public's call for housing justice and tax reform was getting louder, pushing the ruling party and the opposition party to bring these reforms to the agenda of the 2016 presidential election.

Over the past decade, residents in the capital city, Taipei, have witnessed the largest housing price bubble and suffered the toughest problems in housing affordability in the country. Even in the "small government and big market" type of housing market, the city government has made great efforts to establish sound housing market development and to strengthen the rental market. Taipei's experience in housing reform has lessons for other cities in Asia.

This chapter discusses the changes in Taiwan's housing policy and housing wellbeing and their effects on housing justice. The rest of the chapter is organized as follows. The next section discusses the overall changes in housing problems and the change over the past fifty years in the housing policy goal from "helping every dweller to own a home" to "helping every dweller to live in an appropriate home". The third section analyzes the changed policies for delivering housing wellbeing – from the provision of public housing to the provision of mortgage interest subsidies, rental subsidies, and social rented housing in the past four decades. This section also discusses the improvement of housing justice in the last two decades. The fourth section delineates Taipei City's housing reforms to address the worsening problem of high housing prices and to fight speculation in the housing market. The final section discusses the future development of housing market and housing policy. The conclusion also highlights the improvements in housing justice that result from the policy changes.

Housing problem and housing policy: continuity and change

Housing policy in Taiwan has greatly changed over the past several decades. In the late 1950s, housing policy was set to achieve the goal that "those who dwell should be owners" (Construction and Planning Agency, Ministry of the Interior, 2015). During the late 1950s, housing problems were mainly the shortage of housing supply in urban areas, poor living environments and high house prices in main cities (Hsia, 1988). The general public viewed these problems as arising from an imbalance between housing demand and supply. As a result, they asked the government to build a large number of low-priced and good quality public housing units to solve the shortage of housing in the housing market.

Since the late 1950s, the government has implemented a series of housing policies focusing on providing loans and public housing units to middle and low-income households. The Public Housing Loans program was initiated to provide loans with below-market interest rates for targeted households to build or buy a home in the late 1950s. Later the housing policy focused on constructing public housing units to solve middle- and low-income households' housing needs under the Public Housing Ordinance enacted in 1975 (Mi, 1988). The amendment of the Public Housing Ordinance in 1982 describes the provision of public housing directly built by the government, by providing loans with below-market interest rates to build or to buy and by encouraging private industries to build and sell to middle- and low-income households (Chen, 1991). Based on the housing goals set in the 1950s, most public housing units were for sale rather than rent to middle- and low-income households.

During the 1980s, a continuous boom in national economy and deregulation in the financial market drove a steady increase in housing prices and a significant rise in the stock market. The rapid rise in housing prices rendered many households unable to buy their own homes. In August 1989, to fight the skyrocketing housing prices in Taipei City and other major cities in the country, the "shell-less snail housing" movement was debuted. The shell-less snails gained their names because people who could not afford sky-high housing prices and rents were categorized as Houseless Persons (Graduate Institute of Building and Planning, 2014). The shell-less snail housing movement was an early milestone of Taiwan's housing movement to urge the government to cope with high housing prices and to make the housing market function better.

As a response to public demand, in the late 1980 the Taiwanese government issued a series of financial strategies, including selective credit control, to slow the overheated real estate and mortgage market. The credit control in the financial and mortgage market had an instant effect: a quick downturn occurred in the housing market, and it lasted more than a decade (Yip & Chang, 2003). During the 1990s, due to the economic distress in the housing market, a large number of vacant public housing units were unsalable. The government declared a delay in investment and encouraged building public housing. Afterwards, the focus of housing policy shifted to promoting mortgages with below-market interest rates[1] to first-time and low-income homebuyers (Chang & Yuan, 2013). At the end of 1999, the government decided to stop directly constructing public housing units.

From the late 1990s to the early 2000s, due to the recession in the housing market, the government provided favorable mortgage subsidy programs to homebuyers in order to promote the housing market. From 1998 to 2006, more than two thousand billion NT dollars[2] of mortgages with below-market rates were provided to young and first-time homebuyers (Construction and Planning Agency, Ministry of the Interior, 2016). The Overall Housing Policy in 2005 amended the housing goal from "helping every dweller to own a home" to "helping every dweller to live in an appropriate home". Since then, the focus of housing policy has been not only to subsidize middle- and low-income homebuyers but also to provide rent subsidies to socially and economically disadvantaged tenants. Starting in 2007, Housing Subsidies for the Youth has provided rent subsidies to young tenants (Ibid.).

Recovering from the outbreak of SARS in late 2003, the housing market started to boom. Housing prices have increased steadily since the mid-2000s.

The global financial crisis caused a significant recession in the housing market but lasted only a few quarters. Housing prices strongly rebounded since the beginning of 2008 and continued to increase in the 2010s (Chang & Chen, 2012). Housing has thus become less affordable for most families in the country. High house prices have been the top discontent in metropolitan areas according to internet polls in 2010 (Hsieh, 2012). As a result, the government initiated a series of financial strategies including a series of selective credit controls in overheated housing transaction areas to stabilize the housing market and to restrain speculation in it (Chang, 2016).

Moreover, the Specifically Selected Goods and Services Tax (the so-called Luxury Tax) was implemented in June 2011 to regulate housing transactions for units owned only a short time. The Luxury Tax levied 15 per cent and 10 per cent of property transaction prices for holding property less than one year and two years, respectively. The implementation of the Luxury Tax and a series of credit controls and financial constrains aimed to restrain roaring housing prices and reduce speculation in the housing market. However, these strategies did not effectively cool the overheated housing market (Wu & Hsieh, 2014).

At the same time, the Housing Act, passed at the end of 2011, is a second milestone in Taiwan's housing policy. The first article of the Housing Act officially declared for the first time that Taiwan's housing goal was "to establish a robust housing market, improve the quality of housing, and thus allow all citizens to enjoy suitable housing and a dignified living environment." The Housing Act also declared the implementation of various housing subsidies to enable socially and economically vulnerable households to live in suitable homes. The Act included interest subsidies on loans for self-built houses, interest subsidies on loans for purchasing houses, interest subsidies on loans for home renovation, rent subsidies, and subsidies on simple home renovation expenses (Article 9, Housing Act, 2017).

In addition to giving housing subsidies, the Housing Act also defines social housing, referring "to housing built by the government or by the private sector with subsidies from the government that is primarily rented and should rent at least 10 per cent to persons with special conditions and identities" (Article 3, Housing Act). This social housing is either directly built by the government or built by private organizations with the government's assistance for land use planning and property taxes. The Housing Act also establishes criteria for suitable housing quality, stabilizes the housing market, and protects equal rights to housing. The passing of the Housing Act is a third landmark in Taiwan's housing policy reform. However, the policy's effect was very limited in restraining high house prices and in reducing speculation in the housing market. The housing prices rose steadily in the early 2010s. A few people profited greatly from the real estate market, but most people faced sky-high housing prices and less affordability. The inequality in incomes and in the housing market worsened. This situation triggered a second housing movement (called the Nest Movement) in October 2014 after the first movement (the so-called Shell-less Snail Movement) in August 1989. In 2014, the Nest Movement campaigned for housing justice and made five demands including constitutionalisation of the right to housing and banning of forced demolition and relocation, increased social rented housing units, implementation of property tax reforms, expansion of the rental housing market while introducing rental regulations, and re-examining public land-related regulations and suspending Affordable Housing projects (Graduate Institute of Building and Planning, 2014).

In the face of Taiwan's 2016 presidential election, the Nest Movement's campaigns successfully pushed the government and the presidential candidates to announce reforms on housing policy and the housing market. The Overall Housing Policy in 2015 declared four principles of the policy reform: to ensure the housing market's wellbeing, to uphold social justice, to encourage non-governmental participation, and to protect the rights of living (Construction and Planning Agency, Ministry of the Interior, 2015). Moreover, the property transaction tax has long been criticized because the tax bases, that is, the land values and house values to be assessed, were evaluated by the public authority, were far below market values and thus benefited many non-resident owners (those with non-owner-occupied units). This tax advantage to non-resident owners has been seen as the most critical problem of housing inequality. In response to requests from scholars and non-profit housing organizations, the government enacted integrated housing and land taxation in the beginning of 2016. The tax on property transactions is now based on actual transaction prices rather than publicly assessed land and house values, thus easing the problem of the low tax base of the property tax. This adjustment can be seen as another milestone of contemporary property tax reform.

During the 2016 presidential election campaign, DPP presidential candidate Tsai Ing-wen announced a series of housing policy reforms including the development of 200,000 social housing units within 8 years, property tax reform and the amendment of the Housing Act. These policy reforms have gradually been implemented after the newly elected President Tsai and her cabinet took office in May 2016. Since then, several thousand units of social housing have been completed. The amendment of the Housing Act was passed in the beginning of 2017. The formulation of regulations regarding the rental market are in progress.

The development of housing wellbeing and improvement in housing justice

Over the past several decades, developments in housing wellbeing and improvements in housing justice in Taiwan as results of the changes in housing policy have varied across the country and over time. Housing wellbeing focuses on housing subsidies and the security of rental housing. The former addresses the shift of housing subsidies from mortgage subsidies to rental subsidies. The latter emphasizes the security of tenants' rights and the promotion of the rental market.

As mentioned earlier, concerns about housing justice arise because of high housing prices and worsening housing affordability. As a result, the government sees the core value of housing justice as "helping every dweller to live in an appropriate home." Based on this core value, housing justice can be interpreted from three perspectives: securing housing rights, improving housing quality and making the housing market function well.

Due to the shortage in the housing supply in the 1960s, housing wellbeing development focused on providing public housing units and also on providing favorable mortgages to low-income households and target groups like laborers and military, public and teaching staff. The outcome of housing wellbeing development was significant. From the 1960s to the 1990s, the government assisted about 404,427 households to reside in public housing units including residence in 174,891 units built by public authorities and 67,479 units built by private organizations, and 120,306 households received favorable mortgage interest subsidies to buy their homes (Construction and Planning Agency, Ministry of the Interior, 2016). Table 6.1 shows the number of public housing units built by the public and the private sectors in the past several decades. From the 1970s to the early 1990s, most public housing units were directly built by the public sector. After the mid-1990s, the government provided substantial incentives to private developers to build public housing units. Since then, private developers gradually took over the government's role as the main public housing supplier until 1999, when the government announced plans to stop housing construction due to the many vacant public housing units on the market.

Table 6.1 Public housing built by the public and private sectors, 1976–1999 (unit: dwelling units)

Type period	*Government built*	*Private built*
1976–1981*	68,347	NA
1982–1985*	26,472	NA
1986–1989*	2,930	502
1990	14,096	562
1991	3,605	6,460
1992	11,424	3,281
1993	5,970	6,074
1994	9,363	9,321
1995	11,092	10,324
1996	9,478	7,575
1997	6,035	6,558
1998	6,017	9,600
1999	88	7,222
Total	174,891	67,479

Source: Construction and Planning Agency, Ministry of the Interior (2016).

Note

* 1976–1981, 1982–1985, 1986–1989 are the three periods when the central government implemented a six-year project, a four-year project and a continuing four-year project for public housing construction, respectively.

After the halt in construction of public housing in 1999, the housing subsidy program shifted to provide favorable mortgage subsidies to young and first-time homebuyers. Due to the recession in the housing market during the late 1990s and early 2000s, the government provided substantial mortgage subsidies to homebuyers in order to heat up the housing market. From 1998 to 2006, more than two thousand billion NT dollars of mortgages with below-market rates were provided to about 950 thousand young and first-time homebuyers (Construction and Planning Agency, Ministry of the Interior, 2016).

The provision of substantial mortgage subsidies and a series of tax deductions in the Estate Tax and the Gift Tax have indeed stimulated the price boom in housing market since the early 2000s. The housing prices steadily soared to more than double in the past decade, especially in major metropolises. Figure 6.1 shows that the national housing price index rose from 50 to 115.2 between the first quarter of 2005 to the first quarter of 2015 and declined a little to 113.5 in the first quarter of 2017. Taipei's housing price index more than doubled during the same period, reached its peak at the first quarter of 2013 and declined until the first quarter of 2017 (Figure 6.1).

The rapid increase in housing prices led to a crucial affordability problem for most households in Taiwan. After the global financial crisis, the average housing price more than doubled while the average household disposable income increased less than 3 per cent. Figure 6.2 shows that the ratio of the median housing price to median annual disposable income rose from 4.3 to 9.5 between the fourth quarter of 2002 and the second quarter of 2017. The mortgage-payment-to-income ratio also rose from 22.3 per cent to 38.9 per cent in the same period (see Figure 6.3), the problem of housing affordability in Taipei, in

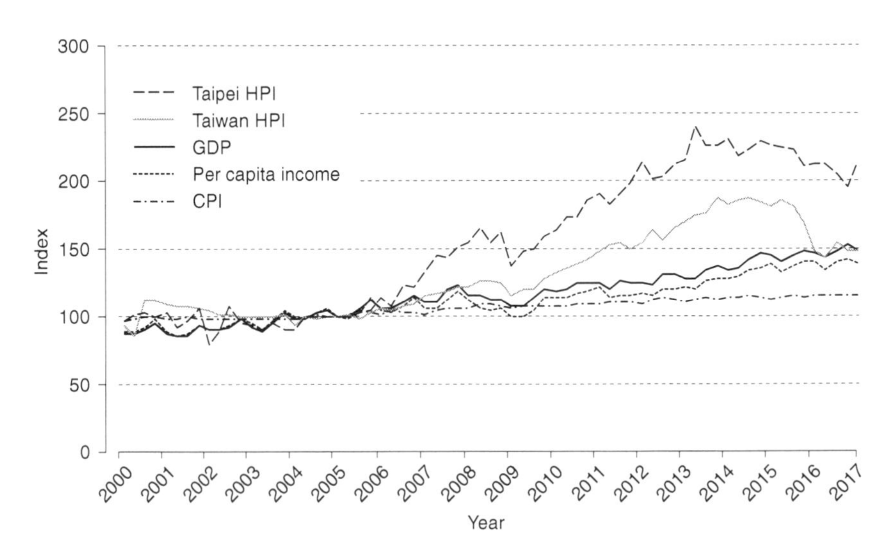

Figure 6.1 House price index and various indexes from 2000 Q1 to 2017 Q1.

Source: Real Estate Information Platform, Ministry of the Interior (various years).

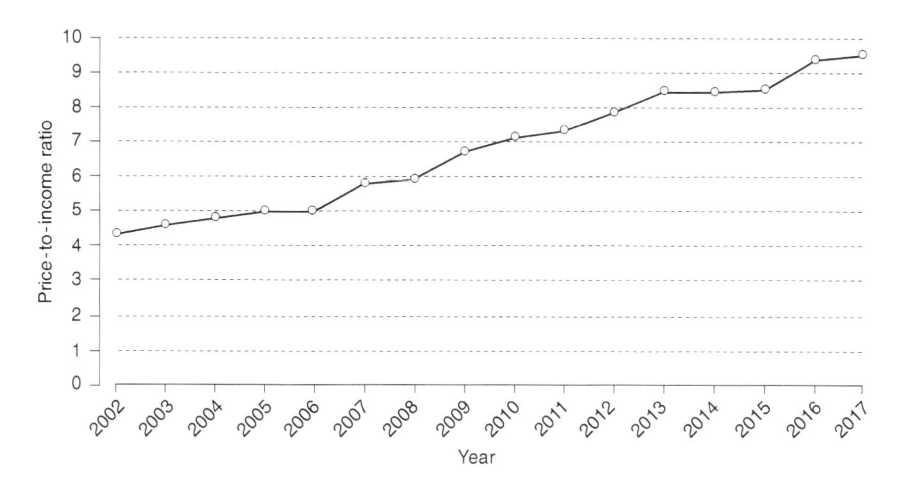

Figure 6.2 The national price-to-income ratio from 2002 Q4 to 2017 Q2.

Source: Real Estate Information Platform, Ministry of the Interior (various years).

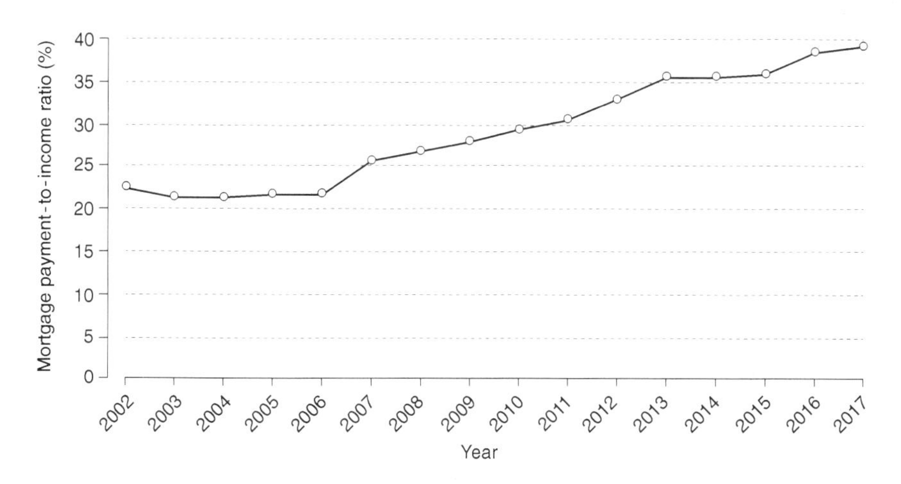

Figure 6.3 The national mortgage payment-to-income ratio from 2002 Q4 to 2017 Q2.

Source: Real Estate Information Platform, Ministry of the Interior (various years).

particular, has worsened greatly. As shown in Figure 6.4, the ratio of median house price to median household disposable income rose from 5.9 to 15.6 between the fourth quarter of 2002 and the second quarter of 2017. The ratio of mortgage payment to median household disposable income also rose from 30.9 per cent to 64.3 per cent in the same period.

The growing discontent with sky-high housing prices and the problem of worsening affordability pushed the government to adjust the focus of its housing policy. Moreover, housing subsidies favoring homebuyers have increased

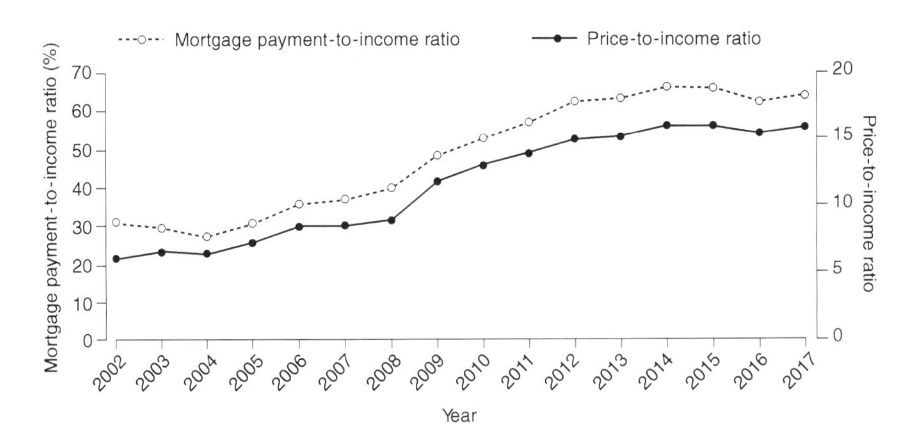

Figure 6.4 The price-to-income ratio and mortgage payment-to-income ratio in Taipei City from 2002 Q4 to 2017 Q2.

Source: Real Estate Information Platform, Ministry of the Interior (various years).

inequality in housing subsidies, which has been criticized by many housing researchers and interest groups (Chang & Chen, 2012; Chen, 2015). As a result, since 2007 the housing subsidy program has focused on three main areas: favorable mortgage subsidies to homebuyers, favorable mortgage subsidies to home owners for home improvement, and rent subsidies to tenants. Figure 6.5 shows the number of subsidized households in the Integrated Housing Subsidy Program from 2010 to 2016.

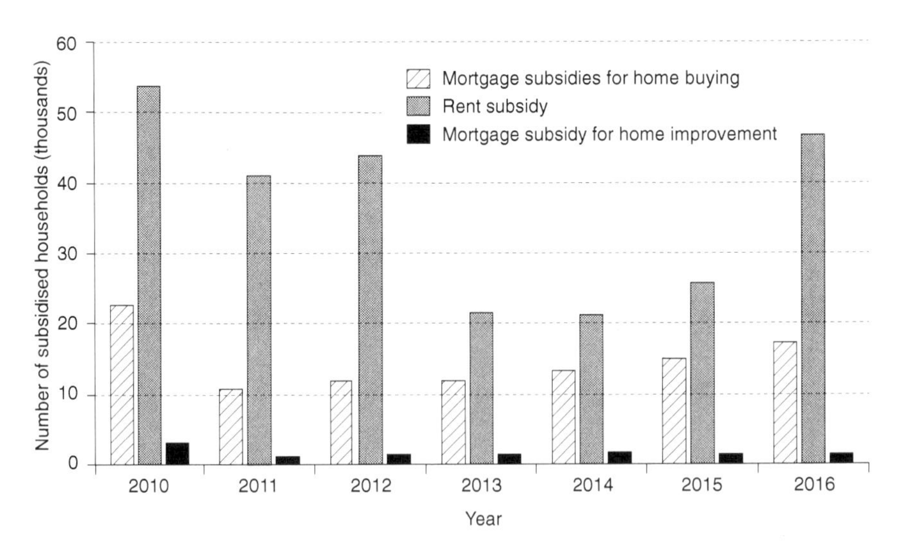

Figure 6.5 Number of subsidized households in integrated housing subsidy program.

Source: Real Estate Information Platform, Ministry of the Interior (various years).

Figure 6.5 shows clearly that the number of households receiving rent subsidies significantly increased in the first few years of the 2010s. This increase occurred because the government allocated much more of its budget to rent subsidies in order to achieve equality in housing subsidies between tenures during this period. Due to limited budgets, the number of rent subsidy recipients decreased sharply in 2013 but has gradually increased since 2014. The number of homebuyers receiving mortgage subsidies reduced significantly from 2010 to 2011. Since then, the number of mortgage subsidy recipients has increased constantly from 12,000 to 17,000 homebuyers.

In addition to vulnerable households, many young households also have suffered serious affordability problems, especially in metropolitan areas since the mid-2000s. As a result, the government also provides subsidies to "young" households–those under 45 years old. The Housing Subsidies for the Youth program includes mortgage subsidies and rent subsidies. Figure 6.6 shows the number of young households who received mortgage and rent subsidies from 2011 to 2014. As the subsidy program is provided mainly to assist young households to buy their own homes, the number of mortgage subsidy recipients increases steadily and is greater than the number of rent subsidy recipients. The rent subsidies in this Housing Subsidies for the Youth program stopped at the end of 2012. Therefore, the number of households that received the two-year rent subsidies reduced sharply by the end of 2014.

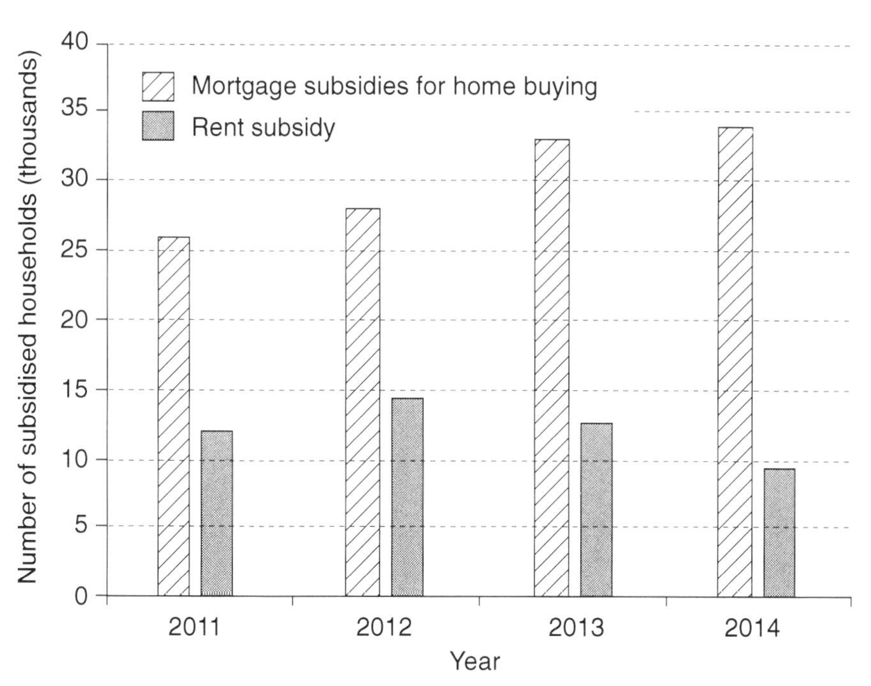

Figure 6.6 Number of subsidized households in housing subsidies for the youth program.

Source: Real Estate Information Platform, Ministry of the Interior (various years).

Taiwan has witnessed significant changes in housing wellbeing over the past several decades. These changes have occurred as the housing subsidy program has shifted from mortgage subsidies to homebuyers to rental subsidies to low-income and vulnerable tenants. However, the security of tenants' rights and the promotion of the rental market have been ignored for a long time until recently. As mentioned above, the new government has made a series of efforts to promote the rental market and to provide security in tenants' rights by enacting the amendment of Housing Act and by drafting the Residential Tenancies Act.

Regarding housing justice, a significant increase in housing quality has occurred in both the public and the private housing sectors over the past several decades. However, inequality in the housing market has worsened. In the past several decades, the housing subsidy policy favored home owners by providing mortgage subsidies and also by selling public housing units to middle- and low-income households, deteriorating equality in housing tenure, especially in the rented sector. As stated earlier, the increasingly high housing prices have worsened the problem of housing affordability for most households in metropolitan areas. For a long time, the property tax favored home owners and damaged tax equality between those with short and longer housing tenures. A typical case of a low holding tax for property is that a household in New Taipei City residing in an average 100-square-meter house pays a total of 2,053 NT dollars for the house tax and the land value tax annually. If the household owns a car with a 2.0 liter engine, the household must pay a total of 17,410 NT dollars for the fuel tax and the license tax annually (Chang, 2016). The low holding tax and transaction tax for property owners have exacerbated inequality in the housing market across the country, but especially in the capital. Taipei has experienced steady progress in the housing market but has suffered the highest housing prices in the past two decades. To cope with these housing problems and to enable robust development in the housing market, the city has made great efforts to implement housing reforms, as discussed in the next section.

Taipei City's housing reform

Taipei, the capital, has experienced a more-than-double increase in housing prices since 2003 and thus made citizens less able to afford their own homes. By the end of 2016, the median housing price-to-median household income ratio reached 15.2, meaning that citizens need to work and save for 15.2 years to buy their own homes without other expenses. The affordability problem in Taipei has become the most critical issue. After the Housing Act passed in 2011, the provision of social rented housing has become an important policy to assist socially and economically vulnerable households. According to the Survey for Social Housing Demand in 2011, the demand for social housing in Taipei City was the greatest in Taiwan. Although there are about 6,800 public rented housing units in Taipei City, rentals account for only 0.73 per cent of the city's total housing stock. The rental housing market regarding the public and the private rented sectors is relatively weak and small, compared to the owner-occupied housing market.

In April 2013, Major Hao Lun-pin had Deputy Major Chang Chin-oh organized two action teams, the Housing Market Action Team and the Rental Housing Action Team, to ensure that the housing market functioned well and to strengthen the rental housing market. The two teams identified several critical problems as well as chaos in the housing market and in the rental market and suggested a series of strategies regarding sound development of the housing market and strengthening of the rental housing market. These are discussed next.

Sound development of the housing market

The Housing Market Action Team identified several critical problems and chaos in the housing market. The first problem was incomplete market information and information asymmetry between real estate developers and homebuyers. Although the Actual Price Registration Scheme was enacted in 2012, it still did not call for disclosure of pre-sale transaction price information. In particular, some media reported unconfirmed real estate transaction information and preferred high-price cases, and thus unscrupulous vendors seize this opportunity to drive up real estate prices. Further, the public sector lacked access to correct and complete information, resulting in an imbalance in housing market information disclosure. In addition, various types of real estate market information are disclosed and interpreted by the real estate industry. This unconfirmed information affects homebuyers' decision-making. The Action Team suggested developing a fair third-sector organization to collect and provide accurate market information, to interpret market information and also to provide assistance to citizens to solve disputes in real estate transactions.

Other problems were overstatement in real estate marketing and advertisements and misleading consumers. Although real estate marketing and advertisement are regulated by the Consumer Protection Act, the Fair Trade Act and the Real Estate Broking Management Act, the real estate authority lacks the manpower to inspect enormous amounts of marketing and advertisement information, which causes many transaction disputes in the real estate market.

Furthermore, the fact that current property taxes cannot fulfill the ability-to-pay principle is a major cause of high housing prices in the city. As stated earlier, the retired Luxury Tax had initially restrained speculation on short-term transactions in the real estate market. Over the long term, a comprehensive real estate tax reform is the effective way to achieve housing justice. To reduce the problem of incorrect and asymmetric information in the real estate market, the Housing Market Action Team suggested that the city government should simplify the registration process and scrutinize the registered transaction prices weekly to filter inappropriate information. As a result, the accuracy rate of transaction price registration has increased from 50 to about 90 per cent per month in recent years (Taipei City Government, 2014a).

To make real estate market information more accurate and transparent, the city government also developed the Taipei City Real Estate Actual Transaction Prices Inquiry System in November 2011 to integrate multiple market information systems into a platform (Taipei City Government, 2014a). The price inquiry system also connects with the Google Map system and the city's library and information center services, which provide more detailed information to sellers and buyers. Moreover, the city government has published the Taipei City Housing Price Index since July 2013. The monthly price index is the first official price index in Taiwan to provide correct and refer market information to citizens and to reduce misleading information from unscrupulous venders. In 2015, the Taipei City Real Estate Integration Information app was developed for citizens to check land, real estate and building information (Ibid.). In June 2013, the city government established the Taipei City Regulation on Real Estate Transaction and Consumption Information Management, providing guidelines for market information announcements by the real estate industry. The city government also inspects market information announced by the real estate industry. Between the second half of 2013 and the first half of 2014, the authority verified more than 1,400 sets of market information (Taipei City Government, 2014a). Moreover, the Taipei City Regulations on Sale of Pre-Sale Housing were established in 2014 to provide guidelines for real estate brokers and marketing agents, to establish order in the sale of property and to enhance security in real estate transactions.

In view of the serious underestimation of the current publicly assessed land values and standard house values, in recent years the city government has dramatically increased the land and house values to narrow the gaps between them and market values. The current assessed land value has increased by 50 per cent since 2013, bringing it to 90 per cent of the market value. This increase would help to reach the goal of achieving the ability-to-pay in the Land Value Tax (Taipei City Government, 2014a). Moreover, the city government has raised the assessed standard house values by 100 per cent to 300 per cent in recent years, thus increasing house taxes up to 160 per cent, which has greatly increased the tax base for the House Tax.[3] Starting in July 2011, the city government also levied additional house taxes for luxury houses in order to reach the goal of fair taxation and ability to pay.

Strengthening rental housing market

In view of the small scale and vulnerable public and private rental markets, the Rental Housing Action Team also delineated some critical and long-term problems in the rental market. In the private rental housing market, most leases are granted by individual landlords or small family enterprises. In particular, much of their rental revenues pays no income tax or business tax. Further, a total of 123 thousand units are vacant in Taipei's housing market, accounting for 13.8 per cent of the total housing stock in 2010 (Directorate-General of Budget, Accounting and Statistics, 2012). The city's vacancy rate is relatively lower than for major cities in Taiwan (see Figure 6.7), but the figure is still higher than for some major cities in Asia. Despite the high vacancy rate, many owners do not want to have their vacant units for lease due to low holding costs and

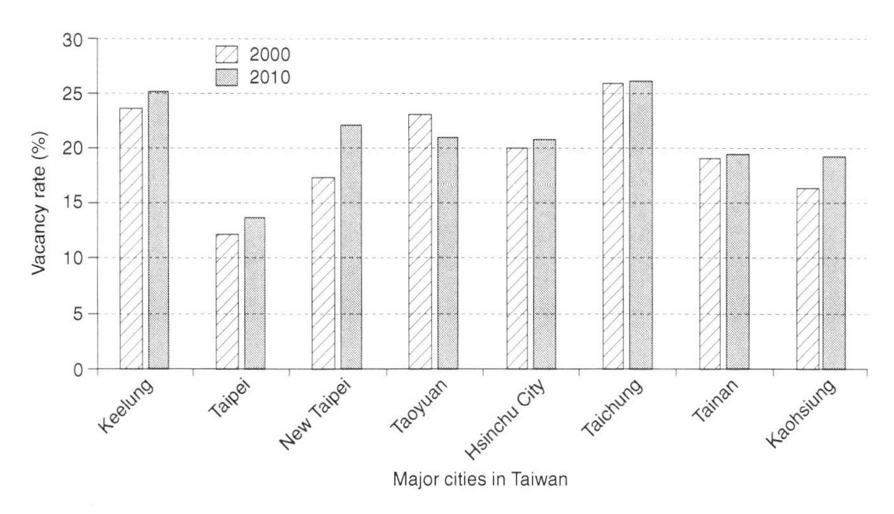

Figure 6.7 Vacancy rates in major cities of Taiwan, 2000 and 2010.

Source: Directorate-General of Budget, Accounting and Statistics, Executive Yuan (2002, 2012).

low rental returns on properties. Also owners expect to have high capital returns from selling their vacant dwellings rather than letting to tenants.

Since most private landlords are individual employers or small family enterprises, many lack professional operation and management knowledge and experience, and they thus gain low returns from renting. The low rental returns also hinder professional and large enterprises from investing in the rental market. In addition, tenancy disputes are the most difficult and complicated affairs to reconcile. Current regulations cannot fairly protect the interests of both landlords and tenants. Therefore, tenancy disputes often take longer to reconcile, causing difficulties for tenancy agreements.

The most critical problem in the public rental market is the shortage of public renting units. There are about 6,000 public rental units, accounting for only 0.6 per cent of the city's total housing stock in 2014 (Taipei City Government, 2014b). The figure is far lower than in some capital cities in Asia. Further, due to a shortage of land suitable for the construction of public rental housing, the supply of public rental housing units is very limited in the city. There is also a lack of sufficient incentives for private organizations to build public rental housing units under the current Housing Act and related regulations. In addition, many public rental housing units were built in the 1950s and 1960s and provided to socially and economically disadvantaged households. As a result, the citizens' general image of public rental housing combines vulnerable households with poor quality. There is an urgent need to improve the citizen's image of public rental housing and its quality.

To solve the shortage of public rental housing units, the city government has employed various ways to increase the number of public rental units. The approaches include regenerating existed old public rental communities, constructing new housing units on public lands, re-allocating housing units from MRT joint development projects, re-allocating housing units from public-led urban regeneration projects, re-allocating housing units from the modification of land use plans and bulk rewards, and providing rewards to private enterprise to build public rental units (Taipei City Government, 2014b). By the end of 2015, about 10,000 public rental units were completed and leased to qualified tenants. Moreover, the city government has spent 380 million NT dollars to renovate interior quality, infrastructure improvement, and soil and water conservation construction in 23 public rental communities since 2012 (Ibid.).

Regarding the problem of rent affordability for low-income and vulnerable tenants, the city government has increased rent subsidies up to 6,000 NTD per month, compared to 4,000 NTD per month in many other cities of the country. The additional 2,000 NT dollars-per-month rental subsidies have benefitted many socially and economically disadvantaged tenants in the city. Since 2012, the city government has spent 462 million NT dollars on additional rental subsidies and has benefitted about 21,000 tenants (Taipei City Government, 2014b).

Moreover, to improve rental service quality and resolve tenancy disputes, in 2013, the city government established the Rental Services Center, which provides rental information, services and tenancy disputes reconciliation. By August 2014, the platform had successfully matched 8 renting cases and reconciled 272 tenancy disputes (Taipei City Government, 2014b). In addition, to encourage private owners to release their vacant houses for lease in the rental market, the city government coordinates with private realtors to establish a private funded renting services platform serving landlords and tenants and also securing tenancy agreements. From September to December of 2013, about 9,600 new cases for lease were offered and about 10 per cent of these cases were rented (Ibid.).

Conclusion

Housing in Taiwan has developed a unique model over the past several decades. The housing problem is quite different from those in other Asian countries. For a long time, the focus of housing policy has been biased toward promoting home ownership, which resulted in a small and vulnerable rental housing market in both the public and the private sectors. The government's long-term lack of intervention in the housing market drove the housing market to a "small government and big market style." The high home ownership rate, high housing prices, and high vacancy rate in the housing market imply a mismatch of housing resource allocation. Faced with these problems and chaos in the housing market, the public's appeals for housing justice and for housing policy and taxation reforms have become the hottest issues on the housing policy agenda. The core value of housing justice, helping every dweller to live in an appropriate home, has become the consensus for most Taiwanese.

All levels of government have made many efforts in housing market reform and tax reform in recent years. In Taipei, the city government has used a series of strategies to ensure sound development of the housing market and to strengthen the rental market. Some strategies, such as providing additional rental subsidies and increasing the holding tax for non-self-occupied owners, have been implemented effectively; some are in process and need continuous observation. Taipei's experience in housing reform can serve as a model to other cities in Asia. However, some strategies, like increasing the tax base of the Land Value Tax and the House Tax and the inspection of real estate projects, have been criticized by many real estate professionals as damaging the housing market rather than saving it. In fact, housing policy and tax reform work neither to damage nor save the housing market. The reforms aim to reach the goals of the housing policy: to establish a robust housing market, to improve the quality of housing and to secure the tenant's right and tenancy agreements. These are also the goals of housing justice.

Over the past several decades, Taiwan has seen a significant improvement in housing quality in both the public and private housing sectors. However, the steady increase in housing prices in metropolitan areas has damaged the affordability of housing for most households. The low holding costs for property owners have exacerbated the inequality in the housing market. The government's long-term neglect of the rental market has caused the small scale of the rental market and a lack of legal protection of the tenant's rights and tenancy agreements. Two housing movements, the Shell-less Snail Movement and the Nest Movement, in the past thirty years have reflected the public's appeals to the government for housing justice. There is still a long way to go to achieve housing justice.

There is also a tough road for housing policy reform in the future. The implementation of integrated housing and land taxation in 2016 can be seen as the third milestone for housing reform. The next step focuses on reforming the Land Value Tax and the House Tax regarding large increases in assessed land values and standard assessed housing values as the tax bases of these two taxes. However, reform of the land value tax and house tax has created much opposition to sharp increases in the tax base of these two taxes. Opponents argue that a rise in tax bases of these two taxes will exacerbate the recession in the housing market.

Moreover, the new amendment of the Housing Act in 2017 has made substantial improvements in current housing policy and subsidy programs. In particular, in addition to new construction, the amendment provides various ways to increase the social rented housing supply such as using public buildings and their sites; renting and managing private buildings for sublease; providing incentives to private organizations to rent and managing private buildings for sublease. There are substantial tax deductions for private landlords who rent to social tenants. Further, the government is drafting the Residential Tenancies Act, which aims to enable robust development of the rental housing market. These strategies will benefit both the public and the private rented housing sectors in the future. There is always a battle for various political parties and interest groups in the policy reform process. Taiwan's housing reform goes its own way but still has a long way to go.

Notes

1 The government started to provide 0.85 per cent below market interest rate to qualified first time homebuyers in 2000, but the interest rate subsidies reduced gradually to 0.125 per cent below the market interest rate in 2004 (Construction and Planning Agency, Ministry of the Interior, 2016).
2 1 TWD = 0.033 USD (as of November 2017).
3 In Taiwan, the holding tax for property is separated into Land Tax and House Tax. The local government levies Land Tax and House Tax annually from property owners. The tax base of Land Tax is publicly assessed land values announced by the public authority. The tax base of House base is publicly assessed standard house values which also are evaluated and announced by the public authority.

References

Chang, C. O. (2016). *Housing justice*. Taipei: Common Wealth Magazine.
Chang, C. O. & Chen, M. C. (2012). Taiwan: Housing bubbles and affordability. In A. Bardhan, R. Edelstein & C. A. Kroll (Eds.) *Global housing market: Crisis, politics, and institutions*. New Jersey: John Wiley & Sons, pp. 447–463.
Chang, C. O. & Yuan, S. M. (2013). Public housing policy in Taiwan. In J. Chen, M. Stephen & Y. Man (Eds.) *The future of public housing: Ongoing trends in the East and the West*. Berlin Heidelberg: Springer-Verlag, pp. 85–101.
Chen, L. C. (1991). Review and evaluation of public housing development in Taiwan, *Proceeding of conference on housing policy and legislation*. Taiwan: Chinese Society of Housing Study.
Chen, Y. L. (2011). New prospects for social rented housing in Taiwan: The role of housing affordability crises and the housing movement. *International Journal of Housing Policy, 11*(3), 305–318.
Chen, Y. L. (2015). The factor and implication for rising housing prices in Taiwan. *Brookings Series, Taiwan-US Quarterly Analysis, 18*, 1–8.
Construction and Planning Agency, Ministry of the Interior. (2015). *Overall housing policy, the 2015 version*. Taipei: Construction and Planning Agency, Ministry of the Interior.
Construction and Planning Agency, Ministry of the Interior. (2016). *Public housing*. Taipei: Construction and Planning Agency, Ministry of the Interior. Available at: https://pip.moi.gov.tw/V2/Z/SCRZ0003.aspx?Func=F_D1&Key=28, accessed on October 4, 2017.
Directorate-General of Budget, Accounting and Statistics, Executive Yuan. (2002). *General report of 2000 population and housing census, Vol. 1 executive summary report*. Taipei: Directorate-General of Budget, Accounting and Statistics, Executive Yuan.
Directorate-General of Budget, Accounting and Statistics, Executive Yuan. (2012). *General report of 2010 population and housing census, Vol. 3 abstract summary report*. Taipei: Directorate-General of Budget, Accounting and Statistics, Executive Yuan.
Directorate-General of Budget, Accounting and Statistics, Executive Yuan. (2017). *Report on the survey of family income and expenditure, 2016*. Taipei: Directorate-General of Budget, Accounting and Statistics, Executive Yuan.

Graduate Institute of Building and Planning. (2014). *1989–2014 Campaigning for urban housing justice: The shell-less snail movement*. Taipei: Graduate Institute of Building and Planning, National Taiwan University. Available at: http://en.bp.ntu.edu.tw/?p=3270, accessed on December 12, 2014.

Hsia, C. J. (1988). *Taiwan's housing policy: The social aspect of public housing program*. Taipei: Department of Civil Engineering, Urban Planning Studio, National Taiwan University.

Hsieh, B. M. (2012). The achievement of housing justice: A review of housing policy and housing act in Taiwan. *Taiwan Environmental and Land Law Journal*, *1*(1), 74–87.

Laws & Regulations Database of The Republic of China. (2017). *Housing Act*. Available at: http://law.moj.gov.tw/Eng/LawClass/LawAll.aspx?PCode=D0070195, accessed on October 22, 2017.

Mi, F. K. (1988). Public housing policy in Taiwan. A Radical Quarterly in Social Studies, *1*(2–3), 97–147.

Taipei City Government. (2014a). *Taipei City's action agenda on the sound development of the real estate market*. Taipei: Taipei City Government.

Taipei City Government. (2014b). *Taipei City's action agenda on rented housing market*. Taipei: Taipei City Government.

Wu, M. C & Hsieh, B. M. (2014). A study on investment preferences of real estate investors after the implementation of specifically selected goods and services tax act – The case of Kaohsiung City. *Chinese Society of Housing Study 2014 Annual Conference paper*. New Taipei City: National Taipei University, December 2014.

Yip, N. M. & Chang, C. O. (2003). Housing in Taiwan: State intervention in a market driven housing system. *The Journal of Comparative Asian Development*, *2*(1), 93–113.

7 No one left homeless

Universal provision of housing in Singapore

Chua Beng Huat and Meisen Wong

> The most important thing we [the PAP government] do for Singaporeans, of course, is to help every family own a home – the HDB flat. The house is much more than a secure roof over their heads. The house in Singapore is also a major way for us to level up the less successful and to give them a valuable asset and a retirement nest egg. We are using the HDB as a means to give every Singaporean household a stake … That's why we are making sure that HDB flats are affordable even to lower-income-households.
>
> (Prime Minister Lee Hsien Long, October 20, 2011)

Introduction

Looking at Singapore today with its obvious economic success in global capitalism, it is easy to forget that the People's Action Party (PAP), which has governed Singapore without disruption since 1959, was a social democratic party. In the widespread post-war anti-colonial, left-infused political atmosphere, the very first local general election was held in 1955, albeit with a limited franchise. Members for an advisory legislature to the British colonial governor were elected by the public. The pro-workers Labour Party won, while conservative political parties of neo-colonialist and wealthy representatives were totally decimated. In 1959, the PAP, then the most radical left-wing social democratic party, won the local election for the first fully elected parliament with overwhelming popular electoral support. The radical faction of the party subsequently left to form the Barisan Sosialis (Socialist Front) and the inheritors of the PAP, under the leadership of Lee Kuan Yew, re-labelled themselves as 'moderates'. Many of the early public policies were grounded in the social democratic ideology: mass education, full employment, public housing, expansion of social welfare and state ownership of part of the economy. During the 1970s, Lee Kuan Yew – convinced that welfare provisions sap work ethics – radically replaced social welfare provisions with a very vehement anti-welfare regime. However, this regime arguably redoubled the emphasis on full employment, which was achieved through low-wage employment in the early phase of the export-oriented industrialization (Rodan, 1989). This industrial strategy is by now the general narrative of the Asian developmental state, of which Singapore is the conventional 'model' for all developing countries, and does not need repeating here (Woo-Cumings, 1999). With reference to the other Asian developmental states of Japan, Korea, Taiwan and Hong Kong, two distinctive features stand out in Singapore: the universal public housing program and a significant state-ownership of domestic and global enterprises, i.e., extensive state-capitalism (Chua, 2016). Both of the distinctive features are critical to the social development of Singapore; the focus here will be on the public housing program.

It is by now a well-worn tale that 90 per cent of the citizens and permanent residents of Singapore live in high-rise flats provided by the Housing and Development Board (HDB), the government public housing authority, and that more than 80 per cent of them own a 99-year lease on the flats in which they reside (Chua, 1997; Wong & Yeh, 1985). This situation would not be possible without the nationalization of land through the draconian acquisition of private land holdings with meagre compensation to the affected landlords. In 1964, the HDB shifted from providing rental housing to selling flats with a 99-year lease. Momentum for the 99-year-lease home ownership picked up when, in 1968, Singaporeans were allowed to draw on their Central Provident Fund (CPF) to finance public housing mortgages. The CPF is a mandatory, tax-exempted social security savings for which the individual saves a proportion of monthly wages, with a proportional monthly contribution from the employer. With this facility, a closed loop of financial transaction for public housing home ownership is established: every month, the home owner's compulsory savings go to the CPF; in turn, the CPF pays a monthly mortgage, on behalf of the home owner, directly to the HDB, the holder of the mortgage. As the monthly savings generally exceed the monthly mortgage, especially in two-income families, home ownership does not affect the owner's existing level of consumption. By the end of the 1990s, approximately 76.5 per cent of Singaporeans and permanent residents "owned" a 99-year-lease HDB flat.

The expansion of public housing home ownership has had profound and varied effects on the social development of Singapore. The obvious benefits are the immediate improvements to housing and environmental conditions, as the entire population was resettled from informal settlements of impermanent building materials–thatched palm-leaf or zinc-sheet roofs, with wood panel siding and other found materials–or over-congested urban shop-houses into comprehensively planned high-rise housing estates. The improved housing environment also improved individual and public health conditions. Less apparent are the effects on employment and wealth creation at individual and national levels. At the individual level, home ownership compels individuals to take on regular formal employment to meet the demand of monthly mortgage payments. Regular employment obviously improves the financial and material conditions of the home owners. The improved housing, environmental and health conditions reduced absenteeism and improved productivity at work. In the early stages of Singapore's economic development, home ownership thus contributed to the transformation of a largely informal labour force of a developing economy into a disciplined industrial labour force. Furthermore, subsidized home ownership enables the government to keep wages relatively low, making Singapore attractive for foreign capital investments.

Home ownership also contributes to capital accumulation, particularly in the past three decades when there has been very rapid housing price inflation. As the public housing flats are paid for by pre-retirement drawing on social security savings, the government has introduced several programs to enable retired home owners to 'monetize' their flats to fund their retirement needs, including the buying back of a portion of the unused 99-year lease for a monthly annuity (Chua, 2015). Institutionally, unlike other Asian developed economies, where the lack of affordable housing stands as an obstacle to marriage and family formation in the face of rapid demographic decline, Singaporeans can readily secure public housing home ownership in preparation for marriage, thus contributing to the stability of the family institution (Teo, 2011). Finally, politically, providing affordable home ownership demonstrates that the PAP is trustworthy in making good its promise to improve the material life of the people, thus contributing greatly to the legitimacy of the incumbent government (Chua, 1997). In sum, the public housing program has indubitably contributed to all levels of the social, economic and political development of Singapore.

Unsurprisingly, as public housing home ownership becomes the norm and an 'entitlement' within the expectations of Singaporeans, the demand to include an ever-increasing proportion of yet-to-be home owners has become an ongoing political issue. In fact, for all but the top 10–15 per cent of income households, the HDB is effectively the only source for housing. The politicised demand to provide housing for all raises two related issues: rapidly rising housing price, on the one hand and, on the other, the financial ability to purchase among the different groups of Singaporeans who still do not own homes. Those potentially marginalized include newly conjugated young families and older households from the low-income strata who have not achieved the financial ability to purchase a flat. For both groups of households, affordable housing is critical to their future.

Affordability

Between 2005 and 2012, the price index for 'resale' public housing flats had risen by 86 per cent; the phrase 'resale flats' colloquially refers to public housing flats placed on sale by the sitting home owners, in contrast to prices of new flats sold directly by the HDB. This rise was the result of a combination of factors. First was a reduced supply of new public housing stock since 2002. Second was a reduced supply of resale flats due to a policy change to incentivise home owners, especially older ones, to rent instead of selling their flats for a life-stream income. Third, economic growth-driven migration policies resulted in a substantial influx of immigrants between 2005 and 2011. Between 2000 and 2010, the non-resident population had risen dramatically to 25.7 per cent of the total population from 18.7 per cent in 2000; between 2005 and 2009, the population of permanent residents also grew at an average of 8.4 per cent annually (Yeoh & Lin, 2012). The massive influx pushed up demand and prices of resale flats, since permanent residents are not entitled to purchase new flats from the HDB. Lastly, prices of resale flats and new ones were intricately entwined in a vicious inflationary cycle (Chua, 2015, p. 10). To maintain the asset value of existing flats, HDB sold new flats at a 20 per cent discount off the prices of equivalent resale flats. This formula created a vicious cycle in which the resale flat prices set the floor for new prices and the prices of new flats in turn led to increases in resale flat prices. The four factors compounded to cause prices of resale flats to spiral upward very quickly (Phang, 2013, pp. 81–82). This upward trend was reflected in a lay person's comment in the local press: "In 1981, I earned S$800[1] plus as a fresh graduate. At that time, one of my colleagues bought a five-room flat for S$35,000. Now, a graduate's pay has risen about four times more but HDB flat prices have risen more than 11 times" (Lee, 2012).

The disproportionate increase in housing prices relative to income is exacerbated by the relative stagnation of wages among the middle class and real income declines in the low-skill, low-income strata. From 1970 to 2000, the period of rapid economic growth, the median monthly income of Singapore labour more than doubled in every decade, with an annual increase from 7 to 9 per cent. After 2000, growth rates for wages slowed to merely 2.5 per cent annually, in spite of continuing national economic growth (Heng, 2013a). Between 2007 and 2013, cumulative inflation rose to a staggering 21 per cent while income increased by 11 per cent; real income declined by 10 per cent (Heng, 2013b). The Economy Society of Singapore further found that between 2001 and 2008, the bottom 30 per cent of the resident working population had experienced a decline in real income, and the next 20 per cent experienced wage stagnation while the income of the top 2 per cent had seen a sharp increase (Bhaskaran, Ho, Low, Tan, Vadaketh & Yeoh, 2013, p. 126). In terms of poverty, in 2010, the Department of Statistics found that 78,641 households were earning less than S$1,500 per month, when the average monthly household expenditure on necessities amounted to S$1,250 (Department of Statistics, 2011). With a Gini co-efficient of 0.46 and above, Singapore's income inequality is one of the highest among the developed economies (Table 7.1).

By the late 2000s, housing affordability was becoming a political issue, in view of the impending general elections in 2011. The foremost opposition party,

Table 7.1 Gini coefficient, 2005–2015

	2005	*2006*	*2007*	*2008*	*2009*	*2010*	*2011*	*2012*	*2013*	*2014*	*2015*
Before taxes and redistribution	0.465	0.470	0.482	0.474	0.471	0.472	0.473	0.478	0.463	0.464	0.463
After taxes and redistribution	0.422	0.418	0.439	0.424	0.422	0.425	0.423	0.432	0.409	0.411	0.410

Source: Department of Statistics, Key Household Income Trends (2015).

the Workers' Party, suggested lowering housing costs by lowering land costs, estimated to constitute approximately 40 per cent of the cost of a new flat, and passing on the savings to consumers (Cheam, 2011). The suggestion was rejected outright by then-Minister of National Development Mah Bow Tan. According to Mah, lowering land value for public housing flats was akin to an "illegal raid" on the reserves because to lower land cost to the HDB would affect the price of state land that the government sells to private developers for housing and commercial developments; the proceeds of these sales are channelled into the national reserves (ibid.). Meanwhile, the progressive unaffordability of public housing for new entrants and low-income households in the housing market began to have negative effects on popular support for the PAP.

The populace's dissatisfaction with housing affordability and immigration, among other grievances, was made evident in the results of the 2011 general elections. The PAP experienced its greatest loss of popular electoral support since 1959, acquiring only 60 per cent of the total votes cast. It lost the highest number of parliamentary seats since 1968: seven, including one of Singapore's most popular cabinet ministers, George Yeo, as a political casualty. The election results compelled the government to carry out significant changes to the economic-growth-at-all-costs approach to governance. In an about-face in wage policy, Acting Minister for Manpower Tan Chuan Jin attributed the slow wage growth for lower-income Singaporeans to competition from foreign labour and proposed stricter regulations on employment passes. The immigration rate dropped dramatically after the 2011 general elections; between 2008 and 2009, the growth rate for non-residents had peaked at 18.8 per cent; in 2012, it fell to 4 per cent; and in 2015, it was a mere 1.9 per cent (population. sg., various years). Additional measures were, however, necessary to cool the housing prices without bursting the bubble.

The government introduced measures to discourage speculative investors and curb demand. In 2013, Additional Buyer's Stamp Duties (ABSD) of 7 and 10 per cent were imposed on Singaporeans for the purchase of second and third properties, respectively, while foreign investors were charged a 15 per cent ABSD, regardless of whether they bought a first or second property.[2] The most radical measure was the implementation for Singapore citizens and permanent residents of a Total Debt Servicing Ratio (TDSR), which stipulates that an individual's total debt servicing ratio – including mortgage, car loan and credit card debt – must not exceed 60 per cent of monthly income. Although the official objective of the TDSR was to institutionalise fiscal discipline on Singaporeans and reduce the risk of bad debt for financial institutions, it had a dampening effect on housing purchases by tightening mortgage availability. In 2011, the Minister of National Development Khaw Boon Wan, delinked the inflationary coupling of prices of new flats to the equivalent resale flats, claiming that the HDB should "be the price-setter, not be the price-follower" and should prioritise "home ownership and affordability" (Chin & Chang, 2013). These cooling measures achieved the 'softening' market effects as intended. Subsequently, housing prices for both public and private housing dropped gradually in each quarter for ten consecutive quarters. In October 2016, prices of private home prices had dipped more than 10 per cent; those of resale public housing flats dropped by approximately 11 per cent, below their last peak in late 2013 (Wong, 2016). Prices appeared to have stabilized. It is uncertain, at the time of writing, how low the government will let property prices decline before removing the cooling measures. With reduced immigration and the decreasing supply of tenants, the rental market has also softened, with declines in rent. The risk margin occurs when prices decline to the point at which they start to threaten the ability of existing home owners to fund their retirement needs from the monetization of their flats. Meanwhile, even with falling prices, households who are currently excluded from home ownership may still be unable to afford the prevailing market prices and require further financial assistance from the government. Consequently, a series of new loan and grant schemes to different categories of would-be home owners has been introduced.

New housing grants for the excluded

Special CPF housing grant

As mentioned earlier, before the 2011 General Election, the government had already begun to address the issue of public housing affordability. One of the measures undertaken was the provision of a Special Housing Grant (SHG) to lower-income families with monthly household income not exceeding S$2,250 of a grant ranging between S$5,000 and S$20,000 in their CPF, which could be used for the specific purpose of purchasing a public housing flat. In addition to the SHG, low-income families would also be entitled to the existing Additional Housing Grant (AHG) that ranges between S$5,000 and S$40,000, graduated according to household income. Families with monthly household income less than S$1,500 would be entitled to the maximum of both grants: of S$20,000 and S$40,000, respectively, a total of S$60,000. It should be noted that singles are also included in this grant scheme. This inclusion is a major policy shift as housing policies in Singapore have conventionally been pro-family and discriminate against singles. However, the demographic and political realities are that the marriage rate has declined significantly, especially among tertiary educated women, who have chosen career over family or, at least, delayed marriage. Their grants are prorated at half the entitlement of a family (Yong & Chin, 2013). With a new two-room flat – the smallest flat for sale – costing between S$80,000 and S$120,000, the combined maximum SHG and AHG grants of S$60,000 would pay for one-half or three-quarters of the cost (Ong, 2011). This assistance should put home ownership within reach for most of the 198,651 resident households, according to the 2010 census, earning below S$1,500 per month, except those whose monthly income is too far below S$1,500, namely the poorest people in the country, among whom are many single-person aged households.

A post-2011 election survey showed that an overwhelming 84 per cent perceived "costs of living" as an important or very important factor of influence in the elections (Institute of Policy Studies, 2011). Obviously, this finding included the middle class, whose wage increases were not commensurate with the inflation of public housing prices. Housing cost is a very significant part of the rising costs of living; housing affordability is thus a cause for concern for the middle class. Unsurprisingly, after the significant decline in electoral support in the election, the government expanded the SHG to embrace middle-income families (Chang, 2013). The grant would also be extended to families with a household income ceiling of S$6,500, therefore including half of Singapore's households (Yong & Chin, 2013). For households with monthly income ranging from S$5,001 to S$6,500, the grant would be graduated from S$5,000 to S$15,000, while families with monthly income of S$5,000 or less would be eligible for the full amount of S$20,000. According to the Minister of National Development, Khaw Boon Wan, this policy was a departure from existing housing policies which have hitherto largely assisted lower-income families (ibid). According to Prime Minister Lee Hsien Loong, the SHG expansion would mean "a family earning S$4,000 a month [could] buy a four-room flat for about S$285,000 and pay only S$67 in cash for their monthly mortgage payments", after the portion that is deducted from their CPF (Chang, 2013).

Silver housing bonus

As self-financing of retirement by individual households is to be derived from monetizing public housing flats, a Silver Housing Bonus (SHB) was introduced in February 2012. The scheme entitles successful elderly applicants to a cash bonus when they choose to 'downsize' their accommodations, from four- or five-room flats to two- or three-room flats, purchased either on the resale market or directly from the HDB. This is to encourage empty-nest families to reduce their housing consumption in order to enhance their retirement fund. To be eligible, an elderly household needs to fall within a monthly household income cap of less than S$12,000. Successful applicants receive a S$20,000 cash bonus: S$15,000 in cash and S$5,000 in their CPF accounts. In addition, applicants must use the proceeds from the sales of the flats to top up their CPF accounts to the minimum sum required for the CPF Retirement Fund: this amounts to S$278,000 for an elderly household (S$139,000 per person). This latter condition made the scheme unattractive because after deducting the cash requirement for the minimum CPF Retirement Fund and a levy on the sale of the flat by the HDB, the cash received from downgrading would be pittance, and the S$20,000 cash bonus was no incentive for the elderly who had relatively high incomes in their pre-retirement careers (Chin, 2012). With public feedback, the Ministry of National Development first reduced the CPF Retirement Fund deduction to S$60,000 cash per household, compared to the original amount of S$139,000 per person; second, the Ministry allowed the subsequent S$100,000 profit from the sale of a large flat to be kept by the household. Finally, any profit in excess of the S$160,000 must be placed in CPF Retirement Accounts. In other words, assuming that profits from sales exceed S$160,000, those who participate in the scheme can acquire only a maximum of S$100,000 in cash, with the rest allocated to compulsory savings. Eligibility rules were also slightly relaxed to include owners of private property with annual rental value of less than S$13,000. Note that this inclusion of elderly private property owners is more symbolic than real, since it is inconceivable in the Singapore housing market to have any private property with such a low annual rental value.

Step-up grant

In contrast to encouraging the elderly to downgrade, a separate grant scheme has been available since 2013 to assist families living in two-room flats whose families have grown to need more living space to upgrade to three-room flats. Successful applicants are granted S$15,000 in their CPFs for the purpose of purchasing three-room flats. Approximately 3,000 families were reportedly eligible for these Step-Up grants. Some conditions of eligibility apply: first, applicants are expected to demonstrate financial stability, defined as continuous employment for 12 months prior to application; second, while the purchase can be either a resale or a newly-constructed flat, both the current and new flats must be located in non-mature estates. The latter constraint minimizes speculation on the part of grant recipients. As flats in mature estates fetch higher prices, the restriction of transaction to immature estates prevents a family from either selling a two-room flat in a mature estate, profiting from buying a three-room flat in a non-mature estate, and still receiving S$15,000 or with the assistance of the S$15,000 grant, upgrading to a three-room resale flat in a mature estate, with the possibility of greater future profit.

Fresh start housing scheme

In early 2016, the government introduced a new housing policy, targeted specifically at those who had been left out of home ownership due to bureaucratic exclusions from housing subsidies. This group constitutes those who were compelled to sell their flats due to divorce or financial circumstances and are now excluded from state subsidies for home ownership since subsidies are provided only to first-time owners. The Fresh Start Housing Scheme (FSHS) aims to provide this group of households, especially those with young children, rental and financial assistance to eventually acquire two-room flats. The rationale for assisting them follows the government's belief in home ownership as a tool of social mobility for low-income households. According to Prime Minister Lee, 'these households often have many different problems – jobs, relationships, children's education, sometimes, drugs. I am very concerned about the future of this group because, without help, they may be permanently out of reach of getting a flat of their own. And they will be trapped in poverty and their children will be affected" (Yeo, 2015). The scheme thus aims at breaking the cycle that reproduces intergenerational poverty.

Eligibility conditions include the following: applicants have to demonstrate financial 'stability', i.e., at least one parent has been regularly employed for at least a year before application; parents must demonstrate fiscal responsibility, defined by whether applicants have incurred less than three months of rental arrears within a year while living in temporary rental accommodation and parents must ensure regular school attendance of their children. Under the FSHS, families receive financial assistance to purchase 2-room flats and must live in the flat for at least 20 years before resale. The lengthy residency requirement prevents families from selling the flat again, to manage financial difficulties. The home ownership grants are disseminated in portions over an extended period. Successful applicants receive a financial grant of over S$30,000; S$20,000 is given to candidates when they receive the keys to their flats, and the remaining amount is distributed to families over a period of 5 years. The continuation in grant disbursement is conditional upon demonstration of financial stability and children's school attendance, to be monitored by officials from the Ministry of Social and Family Development. To date, approximately 1,000 families have qualified for Fresh Start. These conditions of eligibility obviously reflect an embedded assumption of the moral and fiscal irresponsibility of applicants who sell their flats in response to financial austerity. Therefore, there is a 'need' to impose stringent discipline to prevent them from doing so a second time. Many families eligible for the Fresh Start scheme currently live in rental flats under the Interim Rental Housing Scheme (IRHS) and aspire towards home ownership; we say more on rental housing later.

With all the housing grants, government spending on housing assistance began to increase almost immediately after the 2011 election. Table 7.2 shows that the deficit of the HDB rose by 212 per cent in financial year 2011–2012 and continued to rise in subsequent years. Given that HDB is not engaged in any other business activities that might incur dire losses, the rising deficit can be explained only by increased spending on public housing assistance. This phenomenon is reflected in Table 7.2, which shows the significant increases in Additional and Special CPF Housing Grants. In 2013 alone, the government disbursed approximately S$1.1 billion to assist low-income families to purchase their first HDB flats.

Table 7.2 HDB net deficit and CPF housing grant, 2009/2010–2014/2015

	2010–2011	*2011–2012*	*2012–2013*	*2013–2014*	*2014–2015*
HDB net deficit before government grant and taxation (S$ million)	137	427	788	1,978	2,010
Additional and Special CPF Housing Grant received and disbursed (S$ million)	264	439	587	672	753

Source: Compiled and Tabulated from HDB Annual Reports 2009/2010–2014/2015.

Interim rental housing schemes

In spite of the increased subsidies for public housing home ownership, there are still households who remain unable to afford the lowest priced two-room flats. The HDB has divided these into two categories: public rental housing for those who are permanently unable to buy a public housing flat and interim rental housing (IRH) for those who are able to buy and aspire to home ownership. The latter group includes families on the above-mentioned Fresh Start Scheme; the Parenthood Provisional Housing Scheme, which provides temporary rental housing to married or engaged couples and divorced or widowed parents waiting for their new flats from HDB; and families with monthly household income of more than S$1,500 who have fallen on hard times. The 2008 global financial crisis caused 40,000 Singaporeans to lose their jobs between 2008 and 2009 (Ministry of Manpower, 2010). Some of these unemployed would have defaulted on home loans and been compelled to downgrade to rental flats. Temporary rental housing at below-market rates are provided for all three groups as they are deemed able to return to home ownership. The monthly rental rates of unfurnished flats range from S$800 for a three-room flat to S$1,900 for a four-room one, reportedly 40 per cent below prevailing market rates (Chin, 2013).

That temporariness of interim rental housing is inscribed in the conditions of rent and materially in the flats themselves. First, the rental flats are let on a co-sharing basis. Families of supposedly similar social economic backgrounds including religion, race, and family size, as decided by the HDB, are made to share three-room or larger flats, to lower costs of rent and utilities for both families. Not surprisingly, this arrangement regularly results in conflict between the families. Disagreements can occur over the mundane: diets, household chores, splitting of utilities, use of toilet or living spaces. Sofas were not found in some flats because they were a potential source of conflict. MPs described flat-sharing as a source of "continuous problem[s]" (Lim, 2012). Occasionally, the police must be called to quell fights or arguments. Second, notices of rules and regulations are posted within each interim rental unit, including rules advising on the attire to be worn in the flat and reminders not to hoard common spaces in the flat; all aim at maintaining civility and avoiding conflicts between the sharing families. Third, the flats are not equipped with doorbells or mailboxes. The absence of these and other mundane amenities pose inconveniences to the tenants. Frequent inspections by HDB officers also stress the tenants. Finally, flats in the interim rental blocks are also rented, at market prices, to groups of foreign workers and foreign students; these two groups of individuals with limited terms of stay in Singapore reinforce the transient character of the entire housing block (Yeo, 2011). In light of all the features, particularly the mundane conflicts, the social value of co-sharing is questionable, especially for families under the Fresh Start program and newlyweds under the provisional housing scheme, which emphasises welfare and the 'stability' of the living environment for children.

Obviously, the IRHS assumes public housing home ownership as the norm in Singapore. That renting is 'interim' implies that it is a temporary disruption in the expected life trajectory of the tenants. An example is the reported case of Madam Noorliah, an interim rental tenant, who, after a decade of financial difficulty and struggle, finally secured a full-time job after undergoing government-assisted skills retraining. With full-time employment, she signed up for a new three-room flat with the HDB. She speaks of her current accomplishments as a journey teaching that "hard work pays off", reaffirming the work ethic (Basu, 2015a). She is a 'model' for all interim rental tenants. However, how often families in the interim rental flats succeed in purchasing their flats remains an open question. Project 4650 – a community project that provides assistance to households accommodated in two blocks of IRH, reports that between 2012 and 2015, 1000 households lived under the IRH, and approximately a third acquired home ownership (Basu, 2015b). It is unclear whether the two-thirds who remained in the interim arrangement have relocated to the public rental system, where tenancy can be indefinite. Nevertheless, since 2011, an average of 600 households from this segment have purchased their first homes (Heng & Yeo, 2016). This number is a tiny fraction of the 49,162 households in rental flats among 908,499 resident households living in public housing, according to the 2013 HDB Sample Household Survey. Obviously, regardless of how many schemes the government develops to turn Singaporean households into public housing home owners, a significant portion of the population remains too poor to afford the lowest-cost public housing units.

Expansion of public rental housing

To house the lowest income strata, the government has steadily increased the supply of public rental flats in the last decade, through conversion of vacant one- and two-room flats and new construction. This expansion would have been ideologically unacceptable to the late Lee Kuan Yew, who believed that it would produce a "dependency group – those constantly dependent on the government and subsidies" (Chang, 2010). However, economic reality has superseded ideological belief, if homelessness in public is to be avoided. According to the Ministry of National Development, there were 42,000 public rental flats in 2007 and 53,000 in 2016, and the supply is scheduled to increase to 60,000 in 2017 (Koh, 2016). The public rental system of one- or two-room flats, the smallest flats produced by the HDB, constitutes the great majority of public rental housing in Singapore. Eligibility conditions are very restrictive as they are seen as public assistance to the neediest families of Singapore. Only citizens can apply. Monthly household income must be less than S$1,500. Depending on household income and flat size, rent ranges between S$26 and S$275 per month.

The profile of households living in 1- and 2-room HDB flats can be inferred from the 2010 Singapore Census: an overwhelming 73.2 per cent have monthly household income of less than S$1,500; 92.5 per cent have secondary school qualifications or below; less than half have no qualifications at all; an estimated 40.9 per cent are employed in low-wage occupations such as cleaners and labourers (Department of Statistics, 2011). Ethnically, the Malays are overrepresented in the public rental system, as 21 per cent of households versus 13.3 per cent in the total population (Department of Statistics, 2015). In the context of a class-conscious society, public rental flats are laden with stigma. In 2010, some residents in the new towns of Tampines and Pasir Ris protested after it was announced that rental blocks would be built near their homes. A few admitted their concerns over the tenants: 'I don't know what kind of people will be living in the rental block. What if they commit crimes?' (Sudderuddin & Kwong, 2010). Others were more implicit about prejudices, alluding to 'safety' and 'quality of neighbourhood'. There were also complaints about the loss of views due to the raising of the rental block and, consequently, the devaluation of their property (Kok & Ang, 2010). Although residents' claims of possible property devaluation have been denounced by real estate agents as greatly exaggerated, the protests demonstrate the great emphasis residents place on the property values of the public housing flats, the flats being the major source of retirement income for the home owners.

Invisible poverty

Along with the overwhelming majority of Singaporeans who are able to purchase their own flats, the additional schemes of loans and grants have enabled the government to progressively incorporate practically the entire Singapore citizenry into public housing. This move is partly ideological and partly pragmatic. As the government has practically monopolized land ownership, there is no land for alternative housing settlements, other than the very expensive 10–15 per cent of privately held land. The government is thus by default responsible for housing the entire nation except the top income earners, who can afford to live on the 10–15 per cent of private land. The move is ideological because Singapore's image for the world as a very successful economy in global capitalism would be dented if there were widely publicized evidence of abject poverty amidst the apparent prosperity of an essentially middle-class society. No iconic figure better symbolizes abject poverty than homeless individuals or families. Homeless Singaporeans would instantly disrupt the image of success. The late Lee Kuan Yew asked, "You go down New York, Broadway. You will see the beggars ... Where are the beggars in Singapore? Show me. I take pride in that" (Han, Fernandez & Tan, 1998, p. 167). Indeed, when homelessness is reported, the HDB immediately rushes in to provide rental accommodation. Illustrative are the following two cases. A young couple was found to have been living at an airport terminal for seven months (Baharudin, 2015). They apparently declined the offer of accommodation by the Ministry of Social and Family Development; subsequently, their whereabouts have been unknown. The other is a delivery man who lived with his pregnant wife on the deck of his lorry; they were given interim rental housing and have applied to purchase a flat under the abovementioned Fresh Start Scheme (Yeo, 2016). Poverty has been effectively rendered invisible by housing the poor in small rental flats in the public housing estates, where the everyday life of residents is relatively 'homogenous', as income inequalities are not exacerbated by visible differences in housing types (it takes a trained eye to know the difference between a rental block and a home ownership block of housing in a public housing estate), while all receive the same publicly available amenities and services.

Indeed, due to the invisibility of income inequality and poverty in Singapore, poverty did not raise as a public and political issue until the first decade of the twenty-first century. The economy then moved into slow growth, with wage stagnation for the middle class and real decline for low-income workers. Economic inequalities have intensified. Poverty has emerged as a much-publicized and political issue of concern. A quiet encroachment of the destitute and homeless on the streets has also become more visible. Since 2008, family service centres have reported seeing more homeless families camping at the local beaches, stairwells or public housing void decks (Mathi, 2008). The reasons for homelessness vary. Some prefer to eke out a precarious existence on the streets instead of receiving public assistance; some lose family support for different reasons; some have problems maintaining regular employment; and others are ex-home owners who sold their homes and are now ineligible for public housing. Again, after the 2011 general election, the government has become more proactive in addressing poverty, and the above are pragmatic reasons for expanding the public housing program.

In addition to the new housing schemes above, the 2014 budget provided S$8 billion for the Pioneer Generation Package to provide affordable healthcare for senior citizens, who have very low CPF savings because they were employed at the time of very low wages. There are three components to the package – a greater subsidy for outpatient care; injection of money into the healthcare component of the CPF; and life-long subsidies for MediShield Life, a life insurance scheme for senior citizens.[3] The 2015 budget also showed increased social expenditures such as the Wage Credit Scheme, in which the government contributes, for two years, up to 40 per cent of any pay increase for Singaporean workers whose monthly income is less than S$4,000. With increased social expenditures, the 2015 national budget would reportedly incur S$6.7 billion in deficits. However, Low (2015) has argued that the deficit was overstated because certain welfare initiatives that were budgeted for disbursement over an extended period were declared in one year's accounting instead. Regardless, the government has undoubtedly become more liberal in social spending, ideologically emphasising "inclusive growth" and an "inclusive society".[4]

The restriction of immigrant arrival, the inclusion of those yet-to-be home owners and the liberalization of social expenditure have engendered political returns to the incumbent PAP government. In the 2015 general elections, the PAP won 70 per cent of the popular votes, a marked reversal from 60 per cent in the previous general elections. This stunning recovery had also undoubtedly benefitted from the outpouring of emotion in the national mourning of the passing of Lee Kuan Yew, the first Prime Minister, who is revered by Singaporeans for his lifework in developing Singapore, and in the seemingly interminable celebrations of the 50th anniversary of national independence. Significantly, after the 2015 elections, the expansion of social welfare provisions continued.

Showing an ideological influence, the 2015 Budget was entitled "Building our Future, Strengthening Social Security". The Ministry of Finance reported that S$890 million was spent on tax rebates for consumption and CPF top-ups for health expenditure.[5] Public transport fares were reduced by 1.9 per cent, which reportedly would cost public transport providers S$36.1 million of revenue (Siong, 2015). Obviously, the political legitimacy of the PAP has been rejuvenated, and its political dominance continues.

Conclusion

Without doubt, the public housing system has been a critical component of the social and economic development of Singapore. In the first four decades of rapid economic development, the market value of public housing has correspondingly increased, along with the rising income of the populace, giving the PAP government a very high degree of political legitimacy. However, as the Singapore economy – reflecting global market conditions – entered slow growth, stagnated wages, and rising costs of living, the inflationary housing market became a political liability, as identifiable groups of Singaporeans came to be excluded from public housing home ownership. As a consequence, the PAP suffered its greatest losses in both popular votes and parliamentary seats in the 2011 general election. This election setback compelled the government rhetorically and practically towards "inclusive growth", including incorporating more Singaporeans into public housing home ownership and expanding social expenditures and greater social transfers of public resources to needy citizens.

While the measures taken have stemmed the rise of housing prices for now, the primary underlying reason for housing price increase remains: the linkage between public housing home ownership and retirement funding. If the public housing flat is to meet the retirement needs of its owner, its market value must perforce increase annually just to keep up with inflation; otherwise, it would be a negative investment, and its ability to meet the future retirement needs of its owner would be jeopardized. On the other hand, individuals who anticipate their retirement needs must invest in home ownership, thus creating a standing demand for both resale and new public housing flats in spite of the rising prices. However, the government is in no position to break this inflationary tendency by radically reducing the prices of new flats. Although doing so is entirely within its control, such reduced prices would seriously hurt the financial position of all owners of existing public housing flats, in addition to radically destroying the national capital accumulated in the housing stock. The political fallout in terms of popular electoral support and legitimacy to rule would be too much to bear.

Furthermore, while expansion of rental housing for low-income households may have reduced the overt stigma of homelessness due to poverty, it does not address the structural conditions which produce poverty. Wages for low-skill jobs must be increased significantly, and so, too, must social transfers. Official figures for 2015 showed a more optimistic picture. Average income for all workers have experienced real increase; the bottom 10 and 20 per cent of resident households saw a 10.7 per cent and 8.3 per cent real income growth respectively; on average, the bottom 30 per cent of the population saw their average monthly household incomes increase from less than S$1,353 in 2014 to less than S$1,466 in 2015 (Department of Statistics, 2015, p. 8). However, as mentioned above, if S$1,250 monthly is needed for the material necessities for a household to sustain itself, then the bottom 30 per cent of households in Singapore are barely doing so. Also, incidences of homelessness have not decreased. Deeper reforms are therefore necessary to address the rising inequality, to which the inflationary housing market, which is systemically generated by using housing as an asset against retirement, is a significant contributing factor.

Notes

1 1 SGD = 0.74 USD (as of November 2017).

2 Investors from countries which have Free Trade Agreements with Singapore, will be charged Singaporean ABSD rates. This includes foreign states such as the USA, Sweden and Norway.

3 For more information on MediShield life, refer to: www.todayonline.com/commentary/how-medishield-life-benefits-seniors. [Accessed on 7 October 2016].

4 A few notable mentions are as follows: In his 2012 May Day message, Deputy Prime Minister Tharman Shanmagaratnam stated the government mission for "inclusive growth which benefits all Singaporean workers" (Thomas, 2012). The 2013 Budget was titled "A Better Singapore: Quality Growth, An Inclusive Society". At a G20 meeting, PM Lee Hsien Loong stressed the importance of "preparing workers for the future" as crucial to achieving "inclusive growth" in light of the uncertainties in the global economy (Sim, 2015).

5 Information on 2016 Budget available here: www.mof.gov.sg/news-reader/articleid/1661/parentId/59/year/2016/category=Press%20Release. [Accessed on October 7, 2016].

References

Baharudin, H. (2015, 21 November). Trio calls Changi airport home for last seven months. *The New Paper*. Available at: www.tnp.sg/news/singapore-news/trio-calls-changi-airport-home-last-seven-months, accessed on 22 July 2016.

Basu, R. (2015a, 30 August). From heartbreak to hopes of new home. *The Straits Times*.

Basu, R. (2015b, 30 August). Finding the way home; Project aims to help families who are down pick themselves up and be self-reliant. *The Straits Times*.

Bhaskaran, M., Ho, C. S., Low, D., Tan, K. S., Vadaketh, S. & Yeoh, L. K. (2013). Inequality and the need for a new social compact. In S. H. Kang & C. Leong (Eds.) *Singapore Perspectives 2012: Singapore Inclusive: Bridging Divides*. Singapore: World Scientific Publishing, pp. 125–58.

Chang, R. (2010, 28 January). 'No' to more public housing. *The Straits Times*.

Chang, R. (2013, 19 August). Making public housing affordable for all. *The Straits Times*.

Cheam, J. (2011, 26 April). HDB land cost issue still vexes. *The Straits Times*.

Chin, D. (2012, 28 December). More cash for elderly downsizing flat or selling lease. *The Straits Times*.

Chin, D. (2013, 22 January). 7,000 new flats for parents with young kids. *The Straits Times*. Available at: www.straitstimes.com/singapore/7000-new-flats-for-parents-with-young-kids, accessed on 18 May 2016.

Chin, D. & Chang, R. (2013, 13 April). HDB will be the price setter: Khaw. *The Straits Times*. p. A3.

Chua, B. H. (1997). *Political legitimacy and housing: Stakeholding in Singapore*. London: Routledge.

Chua, B. H. (2015). Financialising public housing as an asset for retirement in Singapore. *International Journal of Housing Policy*, *15*(1), 27–43.

Chua, B. H. (2016). State-owned enterprises, state capitalism and social redistribution in Singapore. *The Pacific Review*, *29*(4), 449–522.

Department of Statistics, Singapore. (2011). *Census of population 2010: Households and housing*. Singapore: Ministry of Trade and Industry, Republic of Singapore.

Department of Statistics, Singapore. (2015). *Key household income trends*. Singapore: Ministry of Trade and Industry.

Han, F. K., Fernandez, W. & Tan, S. (1998). *Lee Kuan Yew: The man and his ideas*. Singapore: The Straits Times Press.

Heng, J. (2013a, 6 June). Slower rise in 2012 private-sector pay; real wages dipped because of high inflation; this year likely to be better. *The Straits Times*.

Heng, J. (2013b, 27 April). Entry level pay stagnating because of inflation: experts. *The Straits Times*. p. A2.

Heng, J. & Yeo, S. J. (2016, 7 February). Over 500 rental tenants bought their flats last year. *The Straits Times*.

Institute of Policy Studies (IPS). (2011). *IPS postelection survey 2011*. Singapore: National University of Singapore. Available at: http://lkyspp.nus.edu.sg/ips/wp-content/uploads/sites/2/2013/06/POPS-4_May11_report1.pdf, accessed on 5 October 2017.

Koh, P. K. (2016). Working together for an inclusive and innovative city. *Speeches at 2016 Committee of Supply Debate*. Singapore: Ministry of National Development. Available at: www.reach.gov.sg/~/media/2017/smc/mnd-study-guide.pdf?la=en, accessed on 5 October 2017.

Kok, M. & Ang, Y. (2010, 10 February) Uproar over new rental flats going up. *The Straits Times*.

Lee, C. (2012, 18 October). Housing affordability key to good parenting. *The Straits Times*.

Lee, H. L. (2011). Speech by Prime Minister Lee Hsien Loong at the debate on the president's address, 20 October 2011 at Parliament. *Prime Minister's Office*. 20 October. Available at: www.pmo.gov.sg/newsroom/speech-prime-minister-lee-hsien-loong-debate-presidents-address-20-october-2011, accessed on 6 October 2017.

Lim, D. (2012, 29 July). Not easy when families share flats; scheme helps families with short-term housing, but conflicts lead to calls for review. *The Straits Times*.

Low, D. (2015, 7 March). Budget 2015: In deficit, yet very prudent at heart. *The Straits Times*. Available at: www.straitstimes.com/opinion/budget-2015-in-deficit-yet-very-prudent-at-heart, accessed on 17 August 2016.

Mathi, B. (2008, 22 June). Meet Singapore's nomad families. *The Straits Times*.

Ministry of Manpower. (2010). *Labour market 2009*. Singapore: Ministry of Manpower. Available at: www.mom.gov.sg/newsroom/press-releases/2010/labour-market-2009, accessed on 15 June 2016.

Ong, C. (2011, 4 March). New S$20,000 grant to help needy buy a flat. *The Straits Times*.

Phang, S. Y. (2013). Public housing – appreciating assets. In S. H. Kang & C. Leong (Eds.) *Singapore perspective 2012: Singapore inclusive: bridging divides*. Singapore: World Scientific Publishing, pp. 81–88.

Population.sg. (various years). *Population in brief (various years). Our population, our future.* Available at: www.population.sg/population-trends/demographics, accessed on 5 October 2017.

Rodan, G. (1989). *The political economy of Singapore's industrialization: nation state and international capital.* Basingstoke: Macmillan.

Siong, O. (2015, 23 October). Public transport fares to be reduced by 1.9 per cent from Dec 27. *Channel NewsAsia.* Available at: www.channelnewsasia.com/news/singapore/public-transport-fares-to/2212756.html, accessed on 17 August 2016.

Sim, W. (2015, 16 November). G-20: Preparing workforce for the future is vital to inclusive growth, says PM Lee. *The Straits Times.* Available at: www.straitstimes.com/politics/g-20-preparing-workforce-for-the-future-is-vital-to-inclusive-growth-says-pm-lee, accessed on 19 August 2016.

Sudderuddin, S. & Kwong, D. (2010, February 4). Blocked view is residents' main gripe. *The Straits Times.*

Teo, Y. Y. (2011). *Neoliberal morality in Singapore: How family policies make state and society.* London: Routledge.

Thomas, S. (2012). DPM Tharman on inclusive growth. *AsiaOne.* Available at: http://news.asiaone.com/News/Latest%2BNews/Singapore/Story/A1Story20120427-342330.html, accessed on 19 August 2017.

Wong, A. & Yeh, S. (1985). *Housing a nation: 25 years of public housing in Singapore.* Singapore: Housing and Development Board.

Wong, S. Y. (2016, October 4). Private home prices fall at sharpest pace in 7 years. *The Straits Times.* Available at: www.straitstimes.com/business/property/private-home-prices-fall-at-sharpest-pace-in-7-years, accessed on 6 October 2017.

Woo-Cumings, M. (1999). *The developmental state.* Ithaca: Cornell University Press.

Yeo, H. Y. (2011). *The changing role of hdb rental flats.* Unpublished master's thesis submitted to Department of Architecture, School of Design and Environment, National University of Singapore.

Yeo, S. J. (2015, 24 August). Rental flat boost for families with kids. *The Straits Times.*

Yeo, S. J. (2016, 26 April). Couple living in lorry finally move into flat. *The Straits Times.* Available at: www.straitstimes.com/singapore/housing/couple-living-in-lorry-finally-move-into-flat, accessed on 17 May 2016.

Yeoh, B. & Lin, W. (2012). *Rapid growth in Singapore's immigrant population brings policy challenges. Singapore: Migration policy institute.* Available at: www.migrationpolicy.org/article/rapid-growth-singapores-immigrant-population-brings-policy-challenges, accessed on 1 August 2016.

Yong, C., & Chin, D. (2013, 28 August). Over 50 per cent of households could get housing grant; S$6,500 income gap means more can get help to buy bigger HDB flats. *The Straits Times.*

8 Housing policy in Malaysia

Bridging the affordability gap for medium-income households

Wan Nor Azriyati Wan Abd Aziz, Noor Rosly Hanif, Ainoriza Mohd Aini and Mahazril 'Aini Yaacob

Introduction

The government of Malaysia has long embedded housing as part of the national agenda to provide its citizens with safe, adequate and decent homes. Housing which interconnects with the socio-economic and political environment of the country is acknowledged as a central component of the individual's life. In the context of East Asian Countries, housing is viewed not only as a consumption good, but as an asset to the owner (Ha, 2013). On the other hand, Levitin & Wachter (2013), in their study on the housing bubble in the United States, stated that housing is a unique asset that serves both as a basic consumption good and an investment product.

Housing policy in Malaysia consists of the directions, planning and development of the housing sector at the federal, state and local government levels (Ministry of Urban Wellbeing Housing and Local Government [MUWHLG], 2012, 2013). Powers and responsibilities of each level of government are clearly outlined in the Ninth Schedule and Concurrent Lists in Federal Constitutions. The federal government is responsible for matters relating to regulations and issuance of licensing to the developers while state governments have the power and control over land matters including housing sites. As for the local governments, they have the authority regarding the issuance of development orders and approval of building plans, for which the powers lie under the jurisdictions of the state (Federal Constitution, 2017; MUWHLG, 2013). In terms of the role of the state, it is divided into two categories; one is the federal government, and the other is the state government. The federal government, through MUWHLG, is responsible for formulating the policies and guidelines for housing provisions, whereas the state government is responsible for the implementation of the housing projects in the states.

Continuous efforts are undertaken to ensure that Malaysians of all income levels will have access to adequate, quality and affordable homes. Nevertheless, questions arise as to the challenges faced by both the state and the middle-income households (M40) to create a nation of home owners. This chapter looks into a wide range of programmes that were designed by the government to ensure that affordable homes are available to M40 households and considers their social implications.

Role of federal government

After independence in 1957, there were tremendous changes in the housing policy system, turning from the simple objective of providing housing for government officials only to the objective of providing home ownership opportunities to all citizens, particularly low-income people. The federal government through its national five-year plan has added important components to its development plan (Ha, 2013; MUWHLG, 2013; Shuid, 2015; Zyed, Aziz, Hanif & Tedong, 2014). In the First (1966–1970) and Second National Development Plans (1971–1975), the main provision was for the development of public housing to cater to the needs of the lower-income groups irrespective of their ethnicity (Aziz, Hanif & Singaravello, 2009). The provision for housing was only to promote the welfare of the citizens through the construction of low-cost housing undertaken by the state government with funding from the federal government. The provision of housing for low-income people was further expanded until the Ninth Malaysia Plan (2006–2010). Emphasis was given to provide adequate, affordable and decent housing for all citizens.

The establishment of the National Housing Policy (NHP) in 2011 provided guidelines for the planning and development of the housing sector to all relevant ministries, departments and agencies as well as private sectors. There are three main objectives of this policy: to provide adequate and quality housing with comprehensive facilities and a conducive environment; to enhance the ability of people to own or rent houses; and to set a future direction to ensure the sustainability of the housing sector (MUWHLG, 2012). In line with the above, the federal government was also given the responsibility for providing the grants, loans and technical expertise for housing development. Currently, the federal government is represented by several ministries and government agencies such as the MUWHLG; the Ministry of Federal Territories; the Prime Minister's Office through the 1Malaysia People's Housing programme; the Ministry of Finance through *Syarikat Perumahan Negara Berhad* (SPNB) and also the Ministry of Agriculture.

The role of federal government in affordable housing provision can be seen from the establishment of special agencies such as *Perbadanan Perumahan Rakyat 1Malaysia* (PR1MA) and SPNB. Their establishment clearly changes the structure of housing provision for middle-income people. The state governments instead focus on building affordable housing with cooperation from private developers (Shuid, 2015). Therefore, the Malaysian housing policy has focused on either the direct provision of low-cost housing especially for the lower-income group or subsidising the cost of housing construction for home buyers (M40), particularly those buying homes for the first time (Khazanah Research Institute, 2015).

The Eleventh Malaysia Plan (11MP) (2016–2020) marks the continuation of the housing provision of affordable housing for all. Contrary to the previous Malaysian Plans, the main focus of this 11MP is to serve the basic needs of all population groups irrespective of their income to have inclusive housing provisions. Further, under this plan, the government aims to improve housing affordability and accessibility, especially for the middle- and lower-income groups; strengthening the delivery and management of public housing and promoting an efficient and sustainable housing industry.

Role of the state government

The state government undertakes the key task of identifying state land suitable for the development of affordable houses. The proposed site must also be located at an area which are able to attract potential house buyers. This power is clearly outlined in the Federal Constitution, which confers power to the state for land matters. Therefore, the state is responsible to determine and provide suitable land at strategic locations for housing development. Nevertheless, in reality, Malaysia still suffers from the shortage and unavailability of land; even worse, most of the land in the city centre has been sold to private developers, thus making housing development more expensive. Also, the rising costs of the building materials and labour have resulted in the increase of construction costs such as sand, steel, cement and others. Since the construction of houses relies heavily on manual labour, the construction costs borne by the developers have also increased and hence been translated into the higher prices of houses.

In addition, various state agencies such as Penang Development Corporation, Selangor State Development Corporation and Negeri Sembilan State Development Corporation, among others, have also been involved in housing development. The local governments, on the other hand, are not directly involved in the development of housing but approve planning applications and building plans for both public and private housing projects. Though the local governments do not play any prominent roles in housing development, the scenario is different in the major cities such as the Federal Territories of Kuala Lumpur and Petaling Jaya, mainly due to the fact that the local governments in these cities are directly involved in the administration and maintenance of low-cost public housing (Shuid, 2015). However, the 30 per cent requirement for the allocation of low-cost houses targeted at low-income groups in the housing project developed by private developers (Ariffin, Omar & Sa'dun, 2013) has long been implemented through the local planning authority, which is part of the local government. In Malaysia, private housing developers of projects above certain thresholds in terms of proposed development site size must fulfil this requirement in order to obtain planning permission prior to commencing any development (Abdullahi & Aziz, 2017; Fallahi, 2017).

Home ownership crisis in Malaysia

Several emerging economies including Malaysia share an important common goal of promoting home ownership as a means to drive urban development and urban growth. Housing tenure in Malaysia consists of home ownership (owner-occupancy) and renting (public and private renting) (Hamzah & Adnan, 2016;

Adnan & Hamzah, 2014). Malaysian housing policy, however, favours home ownership instead of other housing tenure (Doling & Omar, 2012; Hamzah & Adnan, 2016; Ronald & Doling, 2010). The home ownership rate was 63 per cent in 2000 and declined to 55 per cent in 2010. Implicitly this decrease suggests that there is an annual decline of about 1 per cent in home ownership. This scenario will thus result in home ownership of less than 50 per cent in 2020 (Olanrewaju, Aziz, Tan, Tat & Mine, 2016). There is substantial evidence that home ownership has created many individual and social benefits. Olanrewaju et al. (2016) further argued that housing in Malaysia is perceived as not affordable if benchmarked against international standards. In general, the cumulative monthly expenditure on rent for all Malaysians is about 27 per cent of household income. However, the percentage is higher (about 35 per cent) for those in low- and middle-income groups. For instance, in 2014, inflation in housing, water, electricity, gas and other fuel categories increased to 3.4 per cent (the rate was 1.7 per cent in 2013). This upward adjustment is reflected in electricity tariffs and a broad-based increase in rentals across different types of residential property. Expenditures on housing (including rent, maintenance, utility bills) are economically substantial, amounting to approximately RM1,600 for a month per household, which accounts for close to 30 per cent of the median income. This percentage means Malaysians spend more of their income on housing than on any other goods and services. However, the concern about housing affordability in urban areas gained substantial momentum in Malaysia as middle-income (M40) households increasingly began experiencing difficulty in accessing home ownership. This M40 denotes households with monthly income ranging between RM3,860 and RM8,320. The 'Budget 2016' mentioned that the definition of M40 would be further reviewed from time to time. However, the Budget 2017 maintained the definition of M40 as per the earlier definition in the Budget 2016. The median income for 2016 (estimated to be RM5,720[1]) demonstrates that this amount is not sufficient for most of the nation to enter into home ownership of affordable housing provided by the government.

The high house prices in urban areas in Malaysia have made it difficult for potential first-time home buyers to purchase houses. From 2012 through 2014, the growth in median house prices outpaced the growth of median household income in Malaysia (Bank Negara Malaysia, 2017) particularly in Kuala Lumpur, Selangor, Penang and Johor (Figure 8.1). New launches have been increasingly skewed towards the higher-end property segments; subsequently, only 18.5 per cent of new housing launches in Malaysia have been priced below RM250,000 (Bank Negara Malaysia, 2017).

On aggregate, houses in Malaysia were considered unaffordable with a median multiple of 4.4 in 2014. Median multiple is also known as the housing-price-to-income ratio is defined as the ratio of the median housing price of the market to the median gross annual household income. The three times median multiple suggests that house prices in the market are subject to minimal distortion, if any, and housing supply is able to meet the effective demand. Research

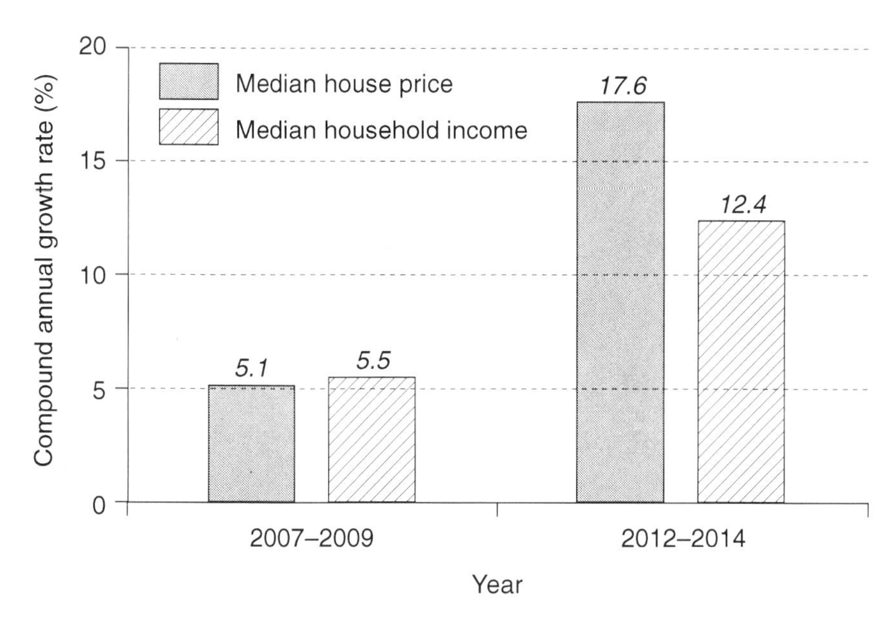

Figure 8.1 Median house price and median household income.

Source: Bank Negara Malaysia (2017).

by the Khazanah Research Institute (2015) based on a median household income of RM55,020 per annum further demonstrated that in the capital city, Kuala Lumpur, housing price was unaffordable as the median multiple reached 5.4. Similarly, another high employment city, Penang, had an unaffordable median multiple of 5.2 (Khazanah Research Institute, 2015).

For many urbanites, living in the city means shorter commutes to workplaces, better amenities and more varied choices of entertainment. With the high cost of living in urban areas, overcrowding and the challenges of an always-on world, it seems surprising that more and more people opt to live in the cities.

The middle 40 per cent of the nation's household income group is extremely important to the nation's wellbeing. The M40 is central to national spending and private investments. A country can reach progressive growth, economically and socially, only with the full participation of the middle class. As such, the Malaysian government has placed greater emphasis on home ownership for all Malaysians, particularly the B40 and M40 groups. Many schemes have been introduced by the public and private sectors to assist these segments of Malaysians. Recognising the issue of upfront payment, the government has included the First House Deposit Financing scheme under the MUWHLG to assist the first-time home buyers to pay for the deposit.

Housing affordability in the Malaysian context

The problem of declining housing affordability has been widely discussed in the media, conferences and even social media. According to the previous study of Aziz et al. (2009), the housing affordability problems of middle-income households in Malaysia are now widespread in most major cities and towns in Malaysia, where many find it difficult to purchase a home. Research by Zyed et al., (2014) demonstrated that escalating housing prices in urban area are the main reason why home ownership is difficult, especially among younger working households.[2] A report by the National Property Information Centre (2016) further illustrates that in 2015, the average housing price was recorded at RM710,089 in Kuala Lumpur and RM467,995 in Selangor. In Quarter 3 (Q3) of 2016, Kuala Lumpur continued to record the highest average housing price (RM772,126), followed by the State of Selangor (RM515,277), which demonstrated a significant increase from the previous year. In addition, a study by Baqutaya, Ariffin & Raji (2016) further suggests that respondents voiced concern about having fewer housing choices available to the middle-income group in areas they preferred. Another alarming issue raised by the study pertains to housing loans, which are getting stricter.

Previous empirical works regarding potential first-time home buyers in Malaysia have focused on home buyers' preferences (Tan, 2012) as well as ownership problems. Zyed, Aziz & Hanif (2016) delved further into the home ownership problem among younger working households in Greater Kuala Lumpur and argued that the problem is attributed to high house prices relative to household income and limited choice of affordable housing. This study, however, was limited to younger working households and did not examine other income groups. The earlier study by Sani (2013) assessed housing affordability among the low-income group in Kuala Lumpur using the residual income method. Her work revealed that type of occupation, level of education and number of household members have significant impacts on affordability. Majid, Saud & Daud (2014) made an attempt to link spending patterns among young couples with impact on home-buying decisions. They argued that young couples who own houses are more cautious in their spending compared to non-home owners. Implicitly, this argument suggests that careful financial planning increases chances of home ownership.

Government initiatives in bridging the affordable housing gap for middle-income households

The housing system in Malaysia focuses more on facilitating home ownership than on other housing tenures (Doling & Omar, 2012; Hamzah & Adnan, 2016). Home ownership is seen as vital in Malaysian housing provision and the future development of the country. The recent 2017 National Budget also witnessed the emphasis on housing provisions to the middle-income group (M40). Several programmes of affordable housing were announced to further accelerate and broaden the accessibility of affordable housing to M40 to enable them to enter home ownership and achieve independent living (Ministry of Finance, 2016). In fact, several housing schemes were launched in the past years.

Perumahan Rakyat 1 Malaysia (PR1MA)

Launched in July of 2011, PR1MA, the Malay acronym for the Malaysia People's Housing scheme, was officially launched by the current Prime Minister, Dato Seri Najib Tun Razak, as part of the government initiatives to develop affordable housing for middle-income households. It was established by the government as Perbadanan PR1MA Malaysia under the PR1MA Act 2012. This Act governs the work process in terms of planning, developing, constructing and maintaining high-quality housing that fits the lifestyle concepts for middle-income households in key urban centres (PR1MA, 2017).

Not all the houses developed under PR1MA are classified as low-cost housing. Instead, the development aimed at enabling the middle-income group to have quality affordable homes of market standards (Perbadanan PR1MA, 2015). In 2014, the monthly median income was reported at RM4,585, and the median housing price stood at RM242,000, which exceeds the affordability threshold level of three times median income (Khazanah Research Institute, 2015). As such, to improve housing opportunities among Malaysians, PR1MA offers houses priced between RM100,000 to RM400,000, targeted at Malaysians whose household income ranges from RM2,500 to RM15,000, covering the M40 segment (PR1MA, 2017). A resale restriction was set at ten years.

Despite the government's efforts to assist the younger generation in terms of their housing opportunities, many have reported being unable to secure loans from financial institutions to purchase houses under this scheme (Jamaluddin, Abdullah & Hamdan, 2016). As a result, PR1MA has decided to introduce the Rent-to-Own Scheme to assist young people who are successful ballot applicants unable to secure loans to rent houses for a maximum of ten years before they buy the units (Perbadanan PR1MA, 2015). The scheme, known as a deferred home ownership programme, was initiated to assist applicants to build the financial capability required for loan financing.

PR1MA also launched the introduction of an end-financing scheme for PR1MA houses. This scheme is one of the schemes under the Home Buyer Assistance Programme that was designed for PR1MA home buyers. The packages include financing up to 100 per cent from PR1MA panel banks without down payments required. Implicitly, this package improves the M40 opportunities to enter into home ownership. This effort is a collaboration between the government, Bank Negara Malaysia (BNM), Employees Provident Fund (EPF) and four local banks: AmBank, Maybank, CIMB and RHB. Last but not least, to further expand the housing opportunities for people and to reduce their burden, the government introduced 100 per cent stamp duty exemption for houses priced below RM300,000. This effort was taken to reduce their burden and at the same time stimulate the growth of the housing sector, especially in providing affordable houses. However, as most of the low- and middle-income groups are unable to afford high-price properties and many also suffer from loan rejections due to strict lending guidelines, it results in an increase of overhang high-price housing units in the property market. Thus, housing developers suffered from the low volume of residential transactions and reduced their appetite to re-invest in new housing development, especially for affordable housing. The introduction of the end-financing scheme that allows for 90 to 100 percent stamp duty exemption, thus, can help spur the demand in the housing market as it can improve accessibility to housing, and it indirectly incentivizes developers to build.

Rumah Selangorku

This project was initiated by the state government of Selangor. The key housing policy of the State Government of Selangor (*Dasar Perumahan Negeri Selangor*) is to realise the goal of '*Satu Keluarga Satu Kediaman yang Sempurna*' (One Family One Perfect Home). That is, to ensure that every citizen or family in the State of Selangor will have a good, comfortable and safe residence to live in. The state government has also introduced a new concept of housing: affordable homes that provide more comfort either in terms of size, design or community. *Rumah Selangorku* housing projects aim to give housing opportunities to low- and medium-income groups to own affordable homes. All the houses under these projects were priced between RM42,000 and RM250,000, depending on eligibility based on buyers' monthly household income (Table 8.1). It is noteworthy that houses priced at RM42,000 are targeted at low-income households earning a monthly income of RM3,000 and below.

Rumah Wilayah Persekutuan – RUMAWIP

In the Budget 2013 presentation by the Prime Minister, the government confirmed that it was aware that comfortable and affordable housing is the most basic need of citizens. Therefore, to facilitate the government's aspiration to improve the accessibility to and affordability of housing, the Federal Government Affordable Housing Policy was enacted to enable the Kuala Lumpur City Hall (DBKL), Perbadanan Labuan and Perbadanan Putrajaya to plan and consider planning applications in order to provide affordable housing more effectively. Therefore, the Ministry of Federal Territories and Urban Wellbeing is committed to ensure that all citizens of the federal territory have the opportunity to have their own homes based on the principle of '*Satu Keluarga Satu Rumah*' (One Family One Perfect Home). RUMAWIP was introduced in Wilayah Persekutuan on March 28, 2013 to provide substantial affordable housing. There are also various types of housing to be introduced in the scheme to cater to the varied needs of different income groups in Wilayah Persekutuan,

Table 8.1 Eligibility criteria and various types of *Rumah Selangorku*

Category	*Eligibility criteria*
Nationality	Malaysian
Age	1. 18 years old and above (Type A houses) 2. 25 years old and above (Type B, C & D houses)
Income (RM)	Less than RM3,000/month (Home buyers for type A) RM3,001 – RM10,000/month (Home buyers for Type B, C & D)
House Price (RM)	RM42,000 (Type A houses; 700 sq. ft) RM100,000 (Type B houses; 750 sq. ft) Type C houses: RM150,000 (800 sq. ft) RM180,000 (900 sq. ft) RM200,000 (18' × 60') Type D houses: RM220,000 (1,000 sq. ft) RM250,000 (20' × 60')
Duration/Moratorium	5 years
Category of buyers	1. First-time homebuyers 2. Staying in Selangor 3. For applicant for Type B, C & D houses; must not have owned a house in Selangor. If they have owned a low-cost house more than 5 years, applicant can apply to buy these types of houses subject to the requirement of having sold the low-cost house.
Location of house	Selangor
Participating Banks	All banks
Others	For owner occupation only

Source: LPHS (2017).

ranging from low-cost to medium-cost housing, with prices between RM52,000 and RM300,000 per unit, depending on location (see Table 8.2).

Perumahan Penjawat Awam 1 Malaysia – PPA1M

PPA1M is an affordable home scheme for public servants. PPA1M is one of the government's efforts to help public servants to own quality houses at prices cheaper than market prices. This program was launched in early 2013 following the decision of the Cabinet on January 30, 2013 and March 27, 2013. The cabinet agreed to enable public servants with low and middle incomes to purchase houses, especially in major cities. The objective of PPA1M is to provide an affordable housing scheme to public servants while emphasizing comfort in terms of size, design, quality, location and house prices that are appropriate for them (see Table 8.3). All public servants, including public servants under the

Table 8.2 Rumah Wilayah Persekutuan – RUMAWIP

Category	*Eligibility criteria*
Nationality	Malaysian
Age	21 years old and above
Income (RM)	Less than RM10,000/month (single), Less than RM15,000/month (married)
House Price (RM)	Low-cost house (RM63,000-KL & Putrajaya; RM52,000-Labuan) of 700 sq. ft (3-bedrooms). Medium- to low-cost house (RM63,001 – RM150,000) of 800 sq. ft (3 bedroom) or 650 sq. ft (2-bedrooms and less) Medium-cost house (RM150,001 – RM300,000) of 800 sq. ft (3 bedrooms) or 650 sq. ft (2-bedrooms and less)
Duration/Moratorium	10 years
Category of buyers	1. First-time home buyers. 2. Do not own a house in Wilayah Persekutuan. 3. Working OR staying in Wilayah Persekutuan.
Location of house	Kuala Lumpur, Putrajaya and Labuan
Participating Banks	Not specified
Others	1. For owner occupation only and cannot be rented out without approval from RUMAWIP's supervision agencies (special agencies). 2. Leader of the household can own only a unit of housing in Wilayah Persekutuan. 3. Should the buyers intend to buy a bigger house, the old unit must be sold with the selling price agreed on by the RUMAWIP's special agency.

Source: RUMAWIP (2017).

Table 8.3 Housing for government servants (*Perumahan Penjawat Awam 1 Malaysia – PPA1M)*

Category	*Eligibility criteria*
Nationality	Malaysian
Age	Not specified
Income (RM)	Max RM10,000/month (household income)
House Price (RM)	90,000–300,000
Duration/Moratorium	10 years
Category of buyers	Staff of federal government, state government, statutory bodies (federal or state) and local authorities.
Location of house	Malaysia
Participating Banks	Not specified
Others	Pensioners and contract staff can apply.

Source: PPA1M (2017).

federal government, state government, state/local authorities, statutory bodies and statutory bodies of the federal state, are eligible under the scheme. In this regard, the government has set the target to build 100,000 units of PPA1M by the year 2018.

My First Home Scheme (MFHS)/Skim Rumah Pertamaku (SRP)

In his 2010 Budget Speech, Prime Minister Dato Seri Najib Tun Razak acknowledged the difficulties of young adults who have just joined the workforce. Prior to this, young adults earning less than RM3,000 were neither entitled to buy houses offered in the private market nor eligible for any public housing. Due to the needs of this group and the difficulties faced by them, the My First Home Scheme (MFHS) was first introduced in 2011 to assist this group of people and to reduce the affordability gaps. As government servants are entitled to either government quarters or specific housing schemes designed for them, MFHS is available only to the private sector employees and employees of the statutory bodies for whom staff housing loans are not offered.

The MFHS, known also as *Skim Rumah Pertamaku* (SRP), was introduced by the government to assist young adults to buy their first homes through 100 per cent mortgage loans from financial institutions. This scheme allows home buyers to obtain financing without the need to pay the housing down payment. This effort is seen to be able to increase the home ownership rate among younger generations. Furthermore, this scheme aims to provide opportunities to younger generations to own houses as they have more financing opportunities (Zyed et al., 2014).

This scheme is applicable only for the purchase of residential units located in Malaysia with a property value of RM100,000 to RM500,000. The applications can be made at participating banks through conventional or Islamic banking. In addition, a purchaser needs to self-occupy the house and is prohibited from renting to anybody. In terms of financing, the loan repayment period is up to 35 years or until the borrower reaches the age of 65, when loan repayments can no longer be deducted from salaries. Table 8.4 outlines the criteria for this housing scheme.

The challenges of M40 for home ownership

Several interrelated challenges prevent M40 non-home owners from becoming home owners. The following are the major categories of challenges.

Income barriers

Mohd Aini, Wan Abd Aziz, Hanif & Musa (2017) argued that the income of many M40 households prevents them from being able to afford monthly repayments for standard mortgage loans. Although some potential home buyers' income is sufficient, their high debt commitment would reduce their eligibility

Table 8.4 Eligibility criteria for the application of My First Home Scheme (MFHS)

Category	*Eligibility criteria*
Nationality	Malaysian
Age	Individual up to 40 years
Other criteria	First-time home buyer Single applicant gross income not exceeding RM5,000 per month and joint applicants' gross income not exceeding RM10,000 per month (based on gross maximum income of RM5,000 per month per applicant) Repayment of total financing must not be more than 60% of the net monthly income Must occupy the property
Mode of loan repayment	Financing tenure not exceeding 40 years or owner not over 65 years old Instalment payable via monthly salary deduction Amortising facility only, without redrawable features

Source: *Skim Rumah Pertamaku* (2017).

to obtain the mortgage required for a home purchase. In Malaysia, commercial banks set very rigid requirements for home loans. For instance, Bank Negara stresses that monthly commitments on paying instalments for a house, car and other payments should not exceed one third of the gross monthly household income. These requirements contribute to failures to purchase a home. The situation is even worse for the self-employed as some of them do not have complete financial records or documents though they can afford to own a home.

A recent study by Mohd Aini et al. (2017) provided further information about the financial position of M40 urbanites of Kuala Lumpur. It is estimated the mortgage-to-income ratio (MIR) is at 0.22, i.e., only 22 per cent of the household income can potentially be used to service the mortgage. Affordability is measured by the residual income (income after deducting necessary household expenditures) of potential urbanites plus their savings, assuming they will use these as monthly payments for joining the First Time Home Buyer scheme. The current average price of houses in Kuala Lumpur at Quarter 1, 2017 is RM 760,369. The current PIR is RM 239,195 or approximately 4 times the average annual earning. Clearly, this PIR demonstrates a huge mismatch between housing prices in the current market and the M40 financing capabilities.

Employment status

Other important factors that affect home ownership are employment or occupational status (including self-employment). There is no doubt that employment is important for ensuring that a household has the capability to own a home. In any circumstances, employment reflects the social status of individuals and their ability to own homes. It means that a person must be employed to guarantee that he or she can purchase a property. Lack of employment shows that a person has no stable income to afford home ownership. Most commercial banks look at the income of the potential borrowers before deciding to entertain them. Those who are under contract or temporary employment or still under probation are less likely to get mortgage approval from the local financial institutions, or to secure a loan-to-value of 70 per cent. For these reasons, the home ownership rate has declined as it is often difficult to secure loans. Implicitly this has resulted in an expanding private rental sector in Malaysia.

Down payment or wealth constraints

The main challenge with regard to home ownership is the insufficient funds available for down payment. Due to the rising house prices and slow income growth, prospective home buyers have to offer higher deposits than they would have ten years ago. In Malaysia, Bank Negara (Central Bank) allows purchasers of first and second homes to obtain financing facilities at the LTV ratio at the present LTV level applied by individual banks based on their internal credit policies. Most of the major banks/financial institutions in Malaysia offer up to 90 per cent financing; hence the applicants must use their own savings to pay the upfront costs. Bank Negara imposed a maximum of only 70 per cent LTV for third house financing facilities. Nevertheless, despite the high LTV, many prospective home buyers need to search for alternative sources to meet the full price of house they want to purchase. While a 10 per cent down payment is not demanding, the current prices of new landed and high-rise properties are beyond the reach of a majority of people, especially the M40, as discussed above.

Consumption pattern

Household spending has huge implications for home ownership. A study by Majid et al. (2014) shows that a majority of respondents in a survey representing those unable to buy a house spent much of their incomes on miscellaneous spending, which involves unplanned expenditures such as groceries, dining out, movies, clothing, grooming, women's accessories and makeup while those who already own a house were found to emphasize savings.

Another issue of entering home ownership is unstable financial positions, causing the M40 to defer home ownership. Although existing houses in the market are sometimes cheaper than those newly built and in good locations, a majority of the M40 prefer to purchase newly built houses from the developer over buying from the existing stock. This is mainly because although purchasing a secondhand house sometimes can be cheaper, it requires additional money for renovation work, especially on the toilets and kitchen. The M40 think in shorter timeframes and like the idea of a finished house. Purchasing directly from the developer means securing a deal where they spend less out of pocket as they will get mortgage loans. Purchasing a home from a developer means legal and valuation fees are normally waived, and the developers often offer attractive prices. In recent years, some developers have offered discounts (rebates) as well as a 0 per cent deposit scheme.

Conclusion

Although the house prices in Malaysia, particularly in urban areas, have grown at a faster pace than the income level, ownership remains a high priority for many Malaysians. This chapter has discussed the role of the government in addressing affordable housing issues for M40 in Malaysia. Various programmes have been designed by the federal and local governments to ensure that affordable homes are available to M40 households. Furthermore, a wide range of affordable housing initiatives introduced by the federal and state governments, e.g. PR1MA and SPNB, have made it possible for many M40 households to become home owners. However, the supply has not been able to meet the demand.

The main perception of the M40 group is that house prices are too high for them to afford. Prices of newly launched landed and high-rise properties are beyond the reach of a majority of people, especially the middle-income earners. The M40s' income is argued to be insufficient to afford the monthly repayment of a mortgage for a typical house in urban areas. Although the incomes of some M40s are sufficient to service the monthly mortgage repayment, they may not have sufficient funds available for the 10 per cent down payment.

Against this background, what is needed is a re-evaluation of the ways in which the federal and state governments encourage home ownership. Spending substantially on home purchases is not sustainable over the long term, and the government needs to bridge the gap. What Malaysia needs is a more sustainable strategy and more innovative approaches to enable home ownership. Malaysia should learn from other developed countries like Australia and the US in addressing home purchase challenges. A more focused approach to housing policy is needed to make greater impacts and enhance home ownership in Malaysia because there is a general preference and expectation to own homes.

Notes

1 1 MYR = 0.24 USD (as of November 2017).
2 Zyed et al. (2014) conducted face-to face survey. This is the views of the respondents surveyed. No figure included in the paper.

References

Abdullahi, B. C. & Aziz, W. N. A. W. A. (2017). Pragmatic housing policy in the quest for low-income group housing delivery in Malaysia. *Journal of Design and Built Environment*, 8, 21–38.

Adnan, N. & Hamzah, H. (2014). Provisional attitude of Malaysia Gen Y's towards alternative housing tenure. *International Surveying Research Journal*, 4(2), 21–35.

Ariffin, S. M., Omar, I. & Sa'dun, A. T. (2013). *Senario perumahan:Isu dan amalan [Housing scenario: Issues and practices]*. First Edition. Kuala Lumpur: Dewan Bahasa dan Pustaka.

Aziz, W. A., Hanif, N. R. & Singaravello, K. (2009). A Study on affordable housing within the middle-income households in the major cities and towns in Malaysia. Available at www.inspen.gov.my/inspen/v2/wp-content/uploads/2009/08/Affordable-housing.pdf, accessed on 30 October 2017.

Bank Negara Malaysia. (2017). *Bank Negara annual report 2016*. Kuala Lumpur: Bank Negara Malaysia.

Baqutaya, S., Ariffin, A. S. & Raji, F. (2016). Affordable housing policy: Issues and challenges among middle-income groups. *International Journal of Social Science and Humanity*, 6(6), 433–436.

Doling, J. & Omar, R. (2012). Home ownership and pensions in East Asia: The case of Malaysia. *Journal of Population Ageing*, 5(1), 67–85.

Fallahi, B. (2017). Evaluation of national policy toward providing low cost housing in Malaysia. *International Journal of Social Sciences*, 6(1), 9–19.

Federal Constitution, Malaysia. (2017). www.refworld.org/docid/3ae6b5e40.html, accessed on 30 October 2017.

Ha, S. K. (2013). Housing markets and government intervention in East Asian countries. *International Journal of Urban Sciences*, *17*(1), 32–45.

Hamzah, H. & Adnan, N. (2016). The meaning of home and its implications on alternative tenures: A Malaysian perspective. *Housing, Theory and Society*, 33(3), 305–323.

Jamluddin, N. B, Abdullah, Y. & Hamdan, H. (2016). Encapsulating the delivery of affordable housing: An overview of Malaysian practice. *Paper presented to the* MATEC *Web Conference*, Vol. 66, Article 47.

Khazanah Research Institute (2015). *Making housing affordable*. Kuala Lumpur: Khazanah Research Institute.

Levitin, A. J. & Wachter, S. M. (2013). Why housing? *Housing Policy Debate*, *23*(1), 5–27.

LPHS (2017). *Application for registration of Rumah SelangorKu*. Available at: http://lphs.selangor.gov.my/55-static.html, accessed on 15 October 2017.

Majid, R. A., Said, R. & Daud, N. (2014). The assessment of young couples' behaviour on expenditure towards homeownership. *International Surveying Research Journal*, 4(2), 35–50.

Ministry of Finance. (2016). *2016 Budget*. Available at www.bnm.gov.my/files/Budget_Speech_2016.pdf, accessed on October 30, 2017.

Ministry of Urban Wellbeing, Housing and Local Government (MUWHLG). (2012). *National housing policy*. Kuala Lumpur: Ministry of Housing and Local Government.

Ministry of Urban Wellbeing, Housing and Local Government (MUWHLG). (2013). The national housing policy. In Cagamas (Ed.) *Housing the nations: Policies, issues and prospects*. Malaysia: Cagamas, pp. 107–115.

Mohd Aini, A., Wan Abd Aziz, W. N. A., Hanif, N. R. & Musa, Z. N. (2017). Barricades to homeownership in Kuala Lumpur: Challenges for the urbanites. *Paper presented to International Conference of Applied Economics and Policy*. Kuala Lumpur, 21–22 August.

National Property Information Centre. (2016). *The Malaysian house price index*. NAPIC. Available at http://napic.jpph.gov.my/portal/main, accessed on 30 October 2017.

Olanrewaju, A., Aziz, A. R. A., Tan, S. Y., Tat, L. L. & Mine, N. (2016). Market analysis of housing shortages in Malaysia. *Procedia Engineering*, *164*, 315–322.

Perbadanan PR1MA. (2017). *Official website*. Retrieved from www.pr1ma.com.my, accessed on 15 June 2017.

Perbadanan PR1MA. (2015). *PR1MA Homebuyer Assistance Programme*. Perbadanan PR1MA Malaysia, Selangor.

PPA1M (2017). Housing for Civil Servants Malaysia. Available at www.kwp.gov.my/index.php/en/perkhidmatan/perumahan-ppa1m-rumawip, accessed on 12 October 2017.

Ronald, R. & Doling, J. (2010). Shifting East Asian approaches to home ownership and the housing welfare pillar. *International Journal of Housing Policy*, *10*(3), 233–254.

RUMAWIP (2017). *Policy of Rumah Wilayah Persekutuan*. Available at https://rumawip.kwp.gov.my, accessed on 12 October 2017.

Sani, N. M. (2013). Residual income measure of housing affordability. *International Journal of Advances in Engineering*, *5*(2), 1–8.

Shuid, S. (2015). The housing provision system in Malaysia. *Habitat International*, *54*, 210–223.

Skim Rumah Pertamaku (2017). *Skim Rumah Pertamaku [My First Home Scheme], Cagamas*. Available at http://www.srp.com.my/en/, accessed on 12 October 2017.

Tan, T. H. (2012). Meeting first-time buyers' housing needs and preferences in greater Kuala Lumpur. *Cities*, *29*(6), 389–396.

Zyed, Z. A. S., Aziz, W. A. Azriyati W. N. & Hanif, N. R. (2016). Housing affordability problems among young households. *Journal of Surveying, Construction and Property*, *7*(1), 5–13.

Zyed, Z. A. S., Aziz, W. A. Azriyati W. N., Hanif, N. R. & Tedong, P. A. (2014). Affordable housing schemes: Overcoming homeownership problems. *Open House International*, *39*(4), 5–13.

9 The unfinished agenda

National housing programmes and policy shifts in India

Urmi Sengupta

Introduction

The urban housing market in India is undergoing a rapid transformation. Over the last three decades, the country has witnessed two contradictory trends: a housing boom and simultaneously, a severe housing crisis. This phenomenon has accompanied numerous policy experiments with an aim to address the diverse housing needs of the population. In particular, since 1991, the country has initiated a number of legislative and regulatory reforms embracing a wider ethos of liberalisation, globalisation and neoliberal development. Helped by institutional transformations, India experienced a housing market boom and expanding owner-occupancy, leading to expanding mortgage markets. The first decade of the reform period unleashed unprecedented opportunities for capital accumulation through private housing development with an inordinate emphasis on the 'market model of housing provision', marked by phenomenal growth in satellite townships, speculative activities driving asset prices and, at the same time, rising socio-economic inequalities both within and across cities.

Despite nearly three decades of experiments, the urban housing deficit has persisted, as is glaringly evident in the form of numerous slums and squatter settlements. According to the latest Census (Government of India, 2011), the housing deficit stands at 18.8 million units, and we can expect the situation to worsen as the apparent explosion of the market is marked by the exclusion of the majority of urban poor from the housing market (Sengupta, 2014). This exclusion is particularly problematic as over 95 per cent of the housing need in India stems from the Economically Weaker Sections (EWS) and Low Income households[1] (LIG), representing 56 per cent and 39 per cent, respectively, whereas the Middle Income Group constitutes 4.4 per cent of the total housing need. The current deficit is estimated to grow at a compound annual rate of 6.6 per cent to the year 2022. The unprecedented gap in the market, arising from the unavailability of housing for those in the BPL (Below Poverty Line), EWS and LIG categories, has also meant that the 'market model' of housing has its share of limitations and that housing finance has yet to filter through the lower income segments. This situation underscores the need for promoting 'affordable housing' and 'pro-poor housing' schemes addressing the 'real' housing problem. The scale of housing needs, however, is staggering. With a population of over 1.3 billion, India represents a population equivalent to roughly two Europe in size. The estimated housing deficit (18 million) suggests that a population roughly the size of the entire population in the UK has no access to decent housing and that 1.7 million people are homeless. The country has, in the last decade or so, initiated a number of flagship projects, marking a new phase of government intervention to address the housing needs of the poor.

The purpose of this chapter is to trace the recent policy shifts and continuities by analysing key national housing programmes implemented in the last two decades. For their magnitude, their scale and the amount of subsidy involved, these programmes are considered to be important milestones in re-asserting 'housing' as an issue of national importance for Indian government policy and action. The chapter contends that India's housing transformation remains an unfinished agenda. The recent flagship programmes follow a dualistic approach of social development and economic pursuits. As is evident, no single national programme in India, especially since the advent of neoliberalism, has been developed as a 'housing policy' on its own. There is a duality owing to the twin foci of housing for the low-income segment and a revival of economic growth – the country has focused disproportionately on the latter. The outcome of these programmes is set back by little or no penetration of housing finance into the low-income segments, weak institutional structure and resource base of the implementing bodies and governance gaps in the broad (non)coalition of interests across the spectrum. The following section outlines the advocacies dispensed by international agencies such as UN-Habitat and the World Bank and concurrent shifts in housing approaches in India.

International advocacies and housing approaches

With the institutionalization of housing as an economic vehicle, in the post-war period, international organizations such as UN-Habitat and the World Bank have taken a huge interest in India–first as part of their social responsibility and, later in the post-1991 period, to help India to introduce a series of regulatory, legislative and financial reforms so that Indian housing champions the private housing development model to enter the global capital market. Simultaneously, the enabling housing paradigm was recommended (World Bank, 1993). The advice dispensed in the 1990s was rooted in the twin framework of rescaling the state and enabling the 'market'. The thrust of this framework was to encourage private sector involvement in housing provision, leaving the state to introduce a raft of ideologically driven 'reforms' that would advance privatization and marketization to boost the overall housing supply including low-income housing units. Subsequently, the government has encouraged private, individualised solutions to housing problems. However, lack of access to long-term mortgages for individual households was identified as the major hindrance to access housing from the private market. Meanwhile, over the long term, the World

Bank started developing a housing finance environment in India, which led to the establishment of the National Housing Bank (NHB) in 1987 as an apex housing finance regulator. The last three decades saw a phenomenal focus on housing finance, the rise of housing finance companies and the development of a mortgage system in India. These efforts concentrated mainly on 'middle- and high-income housing' as a target market, commodifying both production and consumption of housing by linking them to the wider economic circuit for profit. The Bank was clear that there is a mutually reinforcing relationship between the production of housing and the accumulation of capital.

The lure of home ownership thus became central to promotion of housing as an investment commodity rather than a social good. Affordable housing as a 'basic right for all' became a mantra because it weaves an unlikely and unfortunate alliance between the real estate industry and home owners and because the 'Indian market seems a fertile source of new demands' (Marcuse, 2012, p. 218). The scope for housing market growth can be conveyed by a single statement – mortgage penetration in India stands at a very low level (8 per cent). Throughout the 1990s, housing finance comprised roughly 50 per cent of the World Bank Group's housing commitments in India in this period. Therefore, unlike the previous decades which witnessed the Bank investing in housing programmes, slum upgrading, sites and services or developing housing policy, in recent decades it changed direction to focus on making housing finance in line with the pervasive neoliberalist principles in the finance and development sectors. However, the World Bank's efforts have been criticised on the grounds that it was not reaching out to the urban poor. The mortgage finance that it helped to set up has an inherent bias against the urban poor (Choguill, 1997; Datta & Jones, 1999; Jones & Ward, 1995; Rahman, 2001; Rolnik, 2013; Smets, 1997), precluding those working in the cash economy and the informal sector. Loans to poor people by banks have many limitations including lack of security and high operating costs. Most importantly, a majority of poor households do not meet the eligibility criteria or in the strict sense comply with the protocols of pedigree set out for accessing formal housing financing.

Recognizing this problem, the World Bank's efforts are now rolling back to target low-income people through microfinance[2] as it sees a huge potential there both as an investment portfolio and as a means of welfare provision. One aspect of such deliberation involves a recognition of the huge market for housing microfinance given that, in terms of market size, India is home to nearly a quarter (roughly 270 million to 450 million) of the world's poor. The Bank aims at creating a twin framework of supporting financial inclusion by providing access to housing finance to low-income and informal sector households; and by strengthening the capacity of financial institutions that target these groups on a market basis. The International Finance Corporation (IFC), the World Bank's investment arm, has started investing in microfinance institutions lately as there is a huge untapped market catering to low-income, rural, and fragile regions. The IFC's tag line has been 'a strong and engaged private sector is indispensable to ending extreme poverty and boosting shared prosperity through unlocking private investment, creating markets and opportunities where they're needed most'.[3] This aim goes alongside building infrastructure and assisting public-private partnerships for the mainstreaming and reforming the investment climate in the country. This is a natural progression leading to the expansion of the market pattern onto a global scale.

The World Bank's principles of economic-enabling have been in general compliant with both the central and state level in India but with a degree of scepticism and caution. In the previous decades, the private sector channelled its attention to deliver middle- and higher-end housing, seeking to shore up its own legitimacy and thus gradually gaining absolute control over the housing market. The global financial crisis in 2008 and the failure of the private sector to supply affordable housing to the poor have been sobering reminders that the private market model of housing provision alone cannot address the housing crisis in India and that 'it works primarily in the interests of the powerful capitalist property sector and not the public' (Marcuse, 2012). In India, housing affordability has been a huge problem historically, but this has intensified in the last decades, particularly between 2003 to 2014, when it became apparent that housing is becoming unaffordable for most households. This sentiment has reverberated globally. A mid-term review of the two decades of the implementation of enabling strategies by UN-Habitat in 2006 (UN-Habitat, 2006) emphasized the global trends of lower housing production, poor targeting and lack of institutional infrastructure. This review did not prompt any slowdown in the implementation of the enablement strategy due to its contextual embeddedness within the popular neoliberalist tradition and associated economic arguments dispensed by major international institutions like the World Bank or Asian Development Bank (ADB). However, the recent urban strategy has an emphasis on an 'inclusive growth'[4] and 'a need for more targeted interventions through measures such as land adjustment, a return to sites and services and subsidies to the poor' (World Bank, 2010). In sum, enabling policies may have worked well in many countries including India, particularly in generating profits for multinationals, financial institutions and huge accumulations of wealth for some urban elites. Nonetheless, they have the weakness of 'grossly increasing the income gap between rich and the poor' (Yap, 2015), deepening inequalities of housing conditions on a scale not seen before. But even if the enabling project was in some sense a failure, it is indisputable that many of its side effects had a powerful and far-reaching impact on India.

In India, the enabling strategies did not meet the housing demand of the poor, but it brought the housing agenda to the spotlight by linking it to the wider capital circuit. By making financing available to the aspiring middle-class households, the enabling policy effectively reconfigured the structure of the housing market and its articulation into the national economy. More fundamentally, it reinforced decentralization and deregulation by shifting the role of government to facilitating and regulating the overall framework within which various actors, particularly the urban poor, can be enabled through various vehicles, particularly microfinance (see subsequent sections). These attempts may have had little effect on the poor as such due to slow scaling-up, but they were powerful effects in themselves. The first wave of microfinance was started as 'not-for-profit organizations' operating under a paradigm that can be classified as "public purpose" organizations with a goal for social development. However, microfinance organizations are moving into the commercial space and 'for profit' and, increasingly, some of the mainstream finance agencies and banks have now entered the microfinance sector. However, it would be important to see if banking interests could keep up the social motive behind the concept of microcredit.

In many South East Asian countries like Singapore, Hong Kong or China, experience shows that slums were eliminated through government plans for public housing. This 'model' might be difficult to replicate in India. However, the South East Asian model shows that without some form of government investment or intervention on the supply side, no country has managed to resolve slum issues in their entirety. Over the past decade, India's government has initiated various programmes involving multi-billion dollar investments, disbursing sophisticated subsidy packages (Sengupta & Tipple, 2007). This could be arguably seen as 'social protection' to complement the spread of the 'market model' of housing that India had embraced in recent years. Also, in government-aided projects, for instance, housing supplied by housing boards or public-private partnership projects, or in the recent 'affordable housing' schemes, it is a mandatory requirement to set aside 25 to 30 per cent output as low-income housing. The general drift of policies is clear. In the past few years, government is incentivising national developers to supply affordable housing. Companies such as the Tata Group's Smart Value Homes and Shapoorji Pallonji Ltd entered affordable housing in a big way with partnerships with the IFC and the ADB in 2015. Clearly, a dominant trend in India shows a re-engagement of the Indian government in low-income housing after two strong decades of pursuing economic development.

As economic interests of various countries are diverse, countries were invariably selective in terms of 'what to adapt' or 'how to interpret the enabling strategies'. For example, fiscal and monetary policies were not the central feature of the UN's enabling housing strategy, but this is the element that was adopted by the governments and policymakers. In a detailed historical study on the UN's ideas, Jolly, Emmerji & Weiss (2009) make a useful distinction among three types of ideas: *positive* – those that are descriptive and based on verifiable evidence; *normative* – those that define how individuals and institutions should behave and what should be achieved; and *causal* – those that concern strategies and means to achieve a given end. The central idea that enabling housing strategies promote is normative: that housing poverty should be eradicated. This is a fine goal, but enabling strategies do not elaborate on the causal ideas about how housing poverty comes to exist and how to end it, which have become the most contentious elements of the pernicious controversies over the strategy until now. Empowerment, community participation and pro-poor growth–the causative ideas–took a back seat in the policy world (see Section 5), whereas financial liberalization, cost recovery, public private partnership (PPP) and the global economy made headway. Policy becomes distorted and social goals become diluted when institutions embrace strategies selectively in their transition years.

Key national housing programmes

Government intervention in housing in different periods of India's recent history is characterized by the object of its emphasis rather than the degree of government involvement. Slums or *bustee*[5] improvement projects took primacy in the period immediately following independence from British rule, bolstered by the ideological drive emanating from the concept of a free India. The central government implemented schemes such as the Subsidised Housing Scheme for Industrial Workers (1952), Low Income Group Housing Scheme (1954), Middle Income Group Housing Scheme (1959) and Slum Clearance and Improvement Scheme (1956). The period between the 1970s and 1990 witnessed a new drive for public housing with the state acting as a developer. The National Fifth Five Year Plan (1974–1979) provided an institutional impetus for the creation of state housing boards to build formal housing for all income groups. Many state governments built public housing, albeit in small numbers, and in many schemes government employees were given priority in allocation. In the 1990s, important changes were taking place in the governance structure that showed the emergence of decentralization and liberalization as the two pillars of governance guided primarily by the World Bank's or UN's demand for private sector-led development or overall fiscal austerity. From the dramatic declaration in the Ninth Plan that "Housing is State [Government's] subject" to decentralisation, through the Seventy-fourth Constitutional Amendment, which made Urban Local Bodies (ULB) responsible for providing services, including housing, in their jurisdictions, the devolved responsibility led to emergence of a particular kind of housing schemes[6] for the poorer section. These schemes were designed by the central government but required matching funds from the state and local level governments and were supposed to be implemented by ULBs. The liberalization in the 1990s led to the paradigmatic shift owing to the market's logic and structural adjustment policies. These include attention to public-private partnership, regulatory reforms and privatisation to boost land supply and deregulate finance mechanisms, firmly establishing the government's role as an enabler of housing, with a focus on all-income housing. The National Housing policy (NHP) of 1994 made explicit recognition of the economic contribution of the housing and construction sectors in generating employment. One of the core elements of the NHP was "to make the state directly responsible for providing housing for the poor", among other things. This was not very different from the 1988 housing policy; however, strategies to achieve the stated aims were clearer, and the document voiced more concern for poorer citizens. Historically, India's government had been loud about its welfare commitments towards housing the low-income population, at least on paper, but 'spending on housing' has been limited. Deficits created by years of underinvestment in affordable housing and rapid urbanization have now started hindering India's global competitiveness.

Tiwari & Parikh (2012) estimate that the total housing shortage in India is approximately 51 million units, roughly three times more than the figure estimated by the 2011 Census, and an additional 113 million houses will be required if semi-permanent units are also replaced. The housing industry in India suggests the need for around 30 million units in all categories to house every Indian, which requires about 300,000 acres of land and half a trillion dollars of investment. This goal would require creating 15 billion square feet of floor space to be built. This is a tall order that has challenges at each stage of the process. The housing shortage among the EWS and LIG is extraordinarily high, with a 96 per cent share of the total shortage. In the last two decades of economic reforms, the private sector has been responding mainly to the demand for urban housing, catering to the middle- and high-income households. The addition of new housing stock in the market has not reduced shortages, implying that the target consumers for the new stock are different from those households who are creating the market demand for housing. The majority of the supply has been absorbed by the growing middle class (roughly 300 million). With a healthy disposable income, it is becoming the biggest consumer group. In India different income groups thus compete to secure new housing, and understandably the poorer group loses out as the stock is unaffordable for the targeted consumer group. In sum, income inequality affects housing opportunities for the poor, and government interventions in housing can help to address this through cross-cutting policy packages targeting investment in housing and promoting more inclusive programmes.

In the following section, some of the recent housing programmes are briefly discussed to elucidate the current policy shifts around approaches to delivery of housing by public bodies.

Pradhanmantri Awas Yojana (PAY) (2015–2022)

Pradhanmantri Awas Yojana (PAY) is the most recent programme of the Prime Minister's scheme, which is essentially a 'Housing for All' scheme. Under this programme the government has pledged to construct up to 20 million houses by the year 2022. It has recently announced that 2,508 cities in 26 states have been selected for PAY. The programme aims to match the current housing deficit of 18.8 million units. According to the Ministry of Housing & Urban Poverty Alleviation (MoHUPA) (2015), PAY aims to provide affordable homes using four delivery approaches (Figure 9.1) covering both supply-side and demand-side subsidies. The first approach relates to *in situ* improvements undertaken using a developer-led redevelopment process selected through an open bidding process. Developers are incentivised through Floor Area Ratio (FAR) relaxation and the Transfer of Development Rights. The second approach relates to enabling the EWS and LIG[7] households through a credit-linked subsidy programme. In this quintessentially demand-driven approach, the government sits

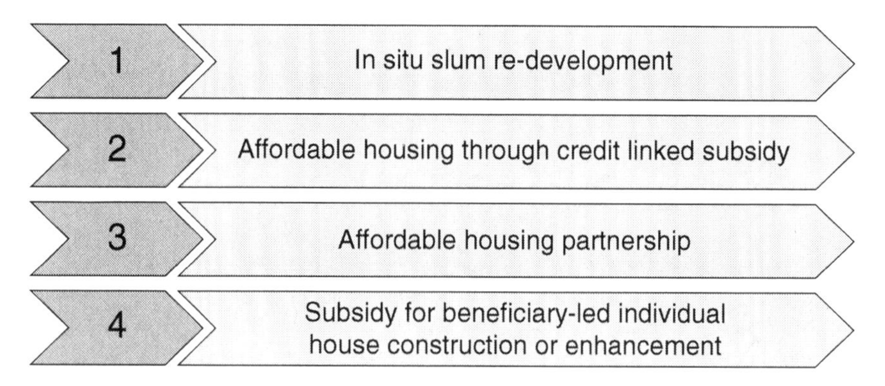

Figure 9.1 The four schemes of PAY.

as a guarantor for disbursement of low-interest loans. Under this programme all eligible households are provided a central grant between 100,000 INR (US$1,490) and 230,000 INR (US$3,400) and loans at 6.5 per cent (4 per cent lower than prevalent housing loans at about 10.5 per cent). The third approach involves a direct subsidy from the central government offered as grant assistance to the public and private sectors (including parastatal agencies) for constructing a minimum of 35 per cent EWS units. The fourth approach provides loans for building individual homes for the EWS population. Collectively, these approaches prioritise women, economically backward groups of Indian society and the scheduled castes and scheduled tribes. Principally, the eligible beneficiaries, most of them slum-dwellers, are classified into the EWS and LIG. According to preliminary estimates, the Housing for All by 2022 scheme will cost the central government about 3 trillion INR (US$45 billion) spread over the next five years.

The programme was launched with self-aggrandizing posters and adverts glorifying the incumbent Prime Minister Narendra Modi for launching a panacea for housing. A close scrutiny of the ingredients of PAY shows apparent ad-hockery and inconsistency. Under all four approaches, 35 per cent output has been reserved for the EWS, and the programme is silent on how the remaining 65 per cent is delivered or distributed across the population. For instance, under the Affordable Housing Partnership Scheme, housing projects can be a mix of different categories of housing, but a project will be eligible for central assistance if at least 35 per cent of the houses are for the EWS category and if the project has over 250 homes (MoHUPA, 2016). Ironically, a scheme promoted as being quintessentially affordable is not wholly targeted to promote home ownership in the EWS category. This situation refutes the principle of positive intervention, since demand coming from EWS far exceeds supply, and in this context, the government should be seeking to fill the gap in housing supply and limit its intervention solely to the EWS/LIG category. In many ways, PAY emulates different genres of housing schemes implemented by the Housing Boards and City Development Authorities in the previous decades. These schemes actively pursued the cross-subsidy approach as a means to deliver public housing units to all income groups.

Another concern is that the 35 per cent of EWS units are distributed using a credit-linked subsidy that falls short of truly estimating the financial capacity of the poor despite being distributed through the Housing and Urban Development Corporation and National Housing Bank, both champions of low-income housing delivery. The subsidy is dispensed at a lower interest rate of 6.5 per cent up to a maximum loan of 600,000 INR (US$9,000) (MoHUPA, 2016). However, despite concerted effort, the credit made available is not conducive to the financial capacity of the poor and in most cases, they are not able to raise the remaining funds. According to the MoHUPA, the EWS and LIG categories typically have annual housing income ceilings of 100,000–300,000 INR (US$1,490–4,500) and 400,000–600,000 INR (US$5,970–8,960) respectively. The benefits are yet to be channelled to the right beneficiaries. While it is early to comment on the performance of the PAY, assumptions that underpin these measures are still guided by the principles of neoliberalism that put profit before people. For instance, the slum upgrading through redevelopment assumes that the excess land that would be available through these projects could be sold on the market instead of being used to improve liveability by creating public spaces or adequate facilities (Patel, 2016), thus restricting the access of slum dwellers to land. This indirect system of governance inevitably embodies a strategy to access the land gradually over time, granting private developers the power to enter, compete in, and encroach on the markets.

These issues have important ramifications. First, the state's mobilizing resources to provide and administer housing delivery that facilitates and structures economic activity is a sign that the government is pursuing economic policy instead of housing policy. Second, market penetration leads to corruption and distortion of the national programme for the benefit of the few. It can be argued that successful implementation of PAY is contingent upon the state's power to restrain the private actors by the rule of law, and this rule of law must be backed by the coercive powers of the state.

Rajiv Awas Yojana (RAY) (2013–2015)

Prior to PAY, the government intervention in slum settlements was marked by another ambitious scheme–Rajiv Awas Yojana (RAY)–launched on September 3, 2013 as a centrally sponsored scheme to deliver slum-free cities. The objective of the programme was to bring existing slums into the formal system with access to basic amenities and to develop institutional and market mechanisms to tackle shortages in land and housing (Tiwari & Rao, 2016). The programme is a reform-linked slum redevelopment and affordable housing programme. RAY envisaged creating social/rental housing, building affordable housing stock in peri-urban areas and undertaking slum rehabilitation projects jointly with the private sector. It was rolled out across 21 cities from 7 states with 55 pilot projects approved for construction of 42,488 dwelling units, including civic and social infrastructure.

Quintessentially a slum rehabilitation scheme with a neoliberal twist, RAY suffered due to both conceptual contradictions and an implementation gap. The RAY model was inherently the one which sought "mainstreaming" of the urban poor through market mechanisms channelled via twin objectives: bringing all existing slums, notified or non-notified (including recognised and identified), within the formal system and enabling them to access the basic amenities that are available to the rest of the city/Urban Authorities. RAY also aimed to redress the failures of the formal system that lie behind the creation of slums by planning for affordable housing stock for the urban poor and initiating crucial policy changes required for facilitating the same. The priority was therefore given to cities with large proportions of slum dwellers. The implementation of RAY is a two-stage strategy that begins with the preparation of a Slum Free City Plan of Action on a 'whole city' basis followed by a Detailed Project Report for slum redevelopment/upgradation/relocation on a 'whole slum' basis. Community participation was considered integral to both stages of work, achieved through close liaison with slum dwellers associations at the slum level and slum dwellers' federations at the city level. The subsidy is shared across the centre and state paying for up to 75–90 per cent of the cost depending on the city's size (Table 9.1). The beneficiaries pay either 25 per cent (for cities larger than 5 million population) or 10 per cent for smaller cities. RAY has been widely and repeatedly depicted as a policy 'failure' by both journalists and scholars. To succeed, RAY required cities to switch to tough policies such as granting property rights to slum dwellers and earmarking 25 per cent of the municipal budget for spending on slum redevelopment. But crucially, unless these are fully and immediately reflected in the states' commitments to prepare the Plan of Action, the programme fails to progress to the next level. Most states expressed reluctance to comply with mandatory provisions for accessing central funds under the scheme and opted out. While the label of 'failure' may be an unfair critique, a closer look at the programme reveals policy biases and gaps. For instance, the programme content overlapped with some of the ongoing urban poor dwelling schemes, the interest of the state has been poorly gauged, and the slum dwellers' apprehension about the idea of relocation has been ignored.

Table 9.1 Funding disbursement in RAY projects

Type of city (2011 census)	*Components*	*Sources of contribution (%)*			
		Centre	*State*	*ULB*	*Beneficiary*
Cities ≥ 50,000	Housing	50	25	–	25
	Infrastructure	50	25	25	–
Cities ≥ 500,000	Housing	75	15	–	10
	Infrastructure	75	15	10	–

Source: MoHUPA (2015).

Jawaharlal Nehru, National Urban Renewal Mission (JNNURM) (2005–2015)

The *Jawaharlal Nehru, National Urban Renewal Mission* (JNNURM) was launched in urban areas in 2005 as an integrated programme aimed at improving housing, water supply, roads, sanitation, infrastructure and transportation in urban areas. Under the JNNURM, the central government gives grants covering 50 per cent of the project cost for cities with populations between one million and four million. For cities with populations higher than four million, the central grant is 35 per cent of the project cost. The remaining funding comes from the state's purse and the urban local bodies or parastatal. At present, there are 523 projects related to urban infrastructure development that are being implemented in 65 cities across the country. JNNURM's largest programme in contemporary India has a two-pronged objective–Urban Infrastructure and Governance (UIG) and Basic Services to the Urban Poor (BSUP), critiques of the programme have asserted that the latter has been allocated far smaller budgetary provisions (25 per cent) than the former, taking the emphasis away from the city's poor. The most significant aspect of this scheme centres on three strands of policy. The first is alignment of urban housing, infrastructure and services (including for the urban poor). Second is developing urban governance through decentralisation in line with the Seventy-fourth Constitutional Amendment Act such that Municipalities (Urban Local Bodies) and the development of new models of low income housing are oriented to high building standards. Third and above all is the scaling-up of housing, which aimed at a much higher production line by rolling out to 65 cities across the nation.[8]

The twin focus of the Mission (funds allocated for 'urban infrastructure' and improvement in basic services for the poor), Mahadevia (2011) argues, creates a situation of deliberate policy confusion. In many ways it reflects the age-old public policy mindset of India that the cities in India need 'infrastructure' and the poor need some consolation/amelioration, whereas in spite of the rhetoric, low-cost housing has not appeared thus far to be a priority for the Mission. For instance, the BSUP have been allocated to only one-fourth of the budgetary provisions, which critics have claimed takes the emphasis away from the city's poor. Another area where JNNURM has not been successful is in the ULB's inability to tap into municipal bonds and in attracting the private sector. It was expected that through reforms, ULBs would be able to access the capital market. But this expectation of leveraging government funds with private finance has remained unfulfilled as most ULBs have not crossed the threshold of reforms that generate a credible revenue model. It is not surprising that the PPPs projects are few and far between. Out of 2,900 urban projects under JNNURM, only about 50 projects had some elements of PPP. There too, the capital investment by private sector was just about 1000 *crore*[9] INR, and none was housing investment (MoHUPA, 2011).

The JNNURM was crafted to transform urban governance and improve services by increasing budgets to cities; however, it simply showed that increasing

Table 9.2 Policies and programmes in India

Date	*Programs/projects*	*Policy visions*	*Interventions*
2015	Pradhanmantri Awas Yojana (PAY)	Housing for All by 2022	20 million homes up to 2022 Beneficiaries mainly EWS and LIG categories in urban areas
2009	Rajiv Awas Yojana	Launched with a vision of a slum-free India on '*whole city approach*'. Aims to create a Mortgage Risk Guarantee Fund to enable provision of credit to EWS and LIG	Central support 75,000 INR per EWS/LIG Dwelling Units (DUs) of size of 21 to 40 sq. m Partnership including private-public A project size of minimum 250 dwelling units is eligible for funding under the scheme. The DUs in the project can be a mix of EWS/LIG-A/LIG-B/Higher Categories/Commercial of which at least 60% of the FAR/FSI is used for dwelling units of carpet area of not more than 60 sq. m
2005 with a 7 year fast-track mission period	Jawaharlal Nehru National Urban Renewal Mission (JNNURM)	Urban focused Reforms and fast track planned development	Rolled out to 65 cities Projects involving Improving basic services of urban poor. Slum improvement and rehabilitation Houses at affordable cost for slum dwellers, urban poor (EWS) (LIG) categories.
1998–99	Indira Awas Yojana	Predominantly rural Women empowerment	2 m (annually) housing programme (35:65 urban rural split) Combination of loan-based/cash subsidy scheme Stricter norms – smokeless kitchens *etc.*
	Slum redevelopment	Subscribe to de Soto (2000) – Slum Real Estate.	

Source: Author.

Note
1 INR = 0.0155 USD (as of November 2017).

funding would not augur well unless local actors are empowered to implement the policy. There has been further criticism as JNNURM has been seen as unabashedly neoliberal in its attempt to secure economic reform in areas including state taxes and with its 'private sector first' approach for all infrastructure. The former is evident from the central government making the grant of JNNURM funds conditional upon the successful reduction of stamp duty, and a series of reforms by the State government that was considered difficult to achieve due to the lack of political appetite. Likewise, the incorporation of slums and *bustees* in the programme was conditional upon formulation of spatial plans that were similarly unachievable for many ULBs that lacked resources. Funding agencies have, however, been insistent on such reforms. For instance, the World Bank, which partly finances an urban reform and management project in Karnataka with JNNURM, threatened to pull out due to the lack of an enabling institutional environment. This threat eventually resulted in the introduction of a number of enabling policy environments including the amendment of legislation (Baindur & Kamath, 2009). In sum, the JNNURM has been designed with a centralized mindset reflecting a top-down planning approach, denying almost any significant role to be played by the state government and urban local bodies, at least in the initial design of the project. These factors have inhibited participation from the states and slowed down the progress. As a result, its performance has been poor: only 26 per cent of the targeted housing has been achieved. Despite putting the right foot forward, the JNNURM buckled under the weight of its own ambitious goal.

The next section presents a more elaborate critique of policy shifts and implementation challenges.

Policy shifts and their implementation

The remaining part of this chapter reflects on the three housing programmes discussed in the preceding section by relating them to the changes in both international policy trends (especially the World Bank's policy shifts) and the domestic focus on low-income housing. It has to be noted that since 1991, rapid growth of the middle-class population and an increased demand for higher-quality housing took the attention away from the low-income sector. Housing finance enabled the upper- and middle-income groups to acquire market housing along with luxury housing in metropolitan cities. According to KPMG (2014), luxury housing had been the fastest-growing market between 2008 and 2012. During just four years, about 182 luxury projects were completed, supplying roughly 25,570 units, each unit costing more than US$170,000 with no cap on the higher side. Within the context provided by the rapid rise of middle-class income and consumption, the government of India has sought to tread a fine line with a much greater degree of pragmatism in its response to international policy trends or that of the World Bank. The booming housing market is seen as the only saving grace for economic growth in the country. The country introduced staged reforms on key real-estate matters (see Sengupta, 2006, 2013, 2014 for details), which have contributed to housing development for middle- and high-income families, which are bolstered by growth in real incomes. Thus, housing investments and the behaviour of house prices not only determined the development of the private real estate market in the past decades but also have the strongest possible influences on the performance of financial intermediaries. India is 'shining', and it is shining for the middle- and high-income population. While incomes of the bottom 50 per cent of the adult population (above 20 years) over the period of 1980–2014 grew at 89 per cent, that of the middle 40 per cent (individuals above the median income and below the top 10 per cent earners) grew by 93 per cent and those of the top 10 per cent grew from 394 per cent to 2,726 per cent (EPW, 2017). Clearly, government intervention in the housing market did not put everyone on the housing ladder but made the inequality worse. The corollary to this view is that the subsidy mechanism built into the public housing programmes is also largely a matter of wider economic concern (Castells, Goh & Kwok, 1990).

However, slow output, coupled with the severity of the housing crisis in (mainly) metropolitan cities put pressure on the central government to change its policies to strongly emphasised affordable housing. Especially with formulation of the National Housing Policy in 2007 followed by the Task Force Report in 2008, affordable housing was formally incorporated into the national policy. Recent programmes such as RAY and PAY were the by-products of the renewed focus on affordability. The JNNURM predated the new focus, but its approaches and ingredients of affordability have easily found a place in this focus. The point is that these programmes reflect a campaign about the need to 'mainstream' affordable housing policies. However, public spending on housing remains a neglected affair despite the fact that building affordable housing on a mass scale is becoming critical for India. The problem is more notional than categorical. From policy documents, it has never been clear whether housing falls under mandatory welfare spending or discretionary spending. Despite the rhetoric, the Indian government has been treating housing as peripheral to other mandatory investments: 'building 200,000 units annually, while 2 million is needed nationwide' (McKinsey Global Institute, 2016). This ambivalence has remained consistent across successive governments. Indian governments have been increasingly discredited for not doing much for the urban poor. Notwithstanding key shifts in the national housing policy, the housing conundrum in India relates to its own complexities and contradictions of its operations, and this is where recent programmes have suffered. While it is early to comment on the performance of more recently announced programmes such as PAY, as of October 2017, 0.2 million (approximately 216,435 units) housing units have been completed under PAY, which constitutes only 0.1 per cent of the target. A further 1.2 million homes (12, 22651) are under-construction which together with completed units makes up roughly 6.2 per cent of the total targeted figure. At this pace, to achieve the target set by PAY, India will have to build 3.6 million homes every year until 2022 to hit the target, which is three times the current figure under construction. Interestingly, great variations exist across different states with the top three progressive states – (Maharastra, Tamilnadu and Guzrat) of India with a GDP share of 30 per cent with 75 per cent output through PAY. The assumptions that underpin the programmes continue to reveal the biases contained with previous programmes, such as JNNURM and the RAY which revolved around "mainstreaming" the urban poor through market mechanisms Kamath (2012) within a decentralised setting. Moreover, each of these programmes is conceptualized at the level of the central government and implemented at local levels separated by their respective focuses, aims and objectives. The former seems to unfold opportunities for the private sector while the latter remains restrictive due to the lack of freedom that lurks in every scheme pertaining to decision-making and institutional capacity.

Challenges in creating conditions for effective operation

The enabling regime of the 1990s envisioned an efficient and privatized urban governance for the delivery of low-income housing that continues to find relevance today. Increasingly, housing poverty in India is internalized into the market mechanism (Sengupta, 2006) through a variety of market-oriented reforms. Enablement of the private sector to provide affordable housing continues to underpin new plans and programmes in India. Efforts have been directed toward increasing the income threshold of the low–income group and removing restrictions on size, thresholds, etc. However, what is problematic is that the covert efforts to attract private developers have been made without creating conditions for the developers to work effectively.

These conditions relate to three main aspects of regulatory reforms, which have been either outstanding or had counterproductive results. First is the slow progress in reforming key legislation such as the Rent Control Act and Urban Land Ceiling Repeal Act (ULCRA) that have historically disincentivised private developers through land supply bottlenecks. These legislations have been found to raise transaction and production costs; as a result, lower profit sectors such as low-income housing projects have suffered. Second, the city-wide master-planning process has been even slower, generating a considerable uncertainty on land availability in the peri-urban area, where new housing development is occurring in most cities. They have also resulted in a general mismatch in supply and demand, leading to vacant units. For example, RAY had produced up to 8,000 dwellings by April 2014, only 1,060 of which were occupied. Low occupancy rates plague similar programs, like Affordable Housing in Partnership (48 per cent) and the JNNURM (Maharashtra alone has over 53,000 unoccupied units built under this scheme) (MoHUPA, 2011). The under-occupation finds its root in the classic mismatch of supply and demand. In most slum development schemes, slum dwellers are relocated at the periphery of cities away from employment and transport nodes as opposed to the centrally located (and therefore more valuable) land they formerly occupied, thereby generating a notion of actually being materially 'worse off'. Third, the reform of the permit system continues to be slow and uneven (Sengupta, 2013). Ram and Needham (2016) argue that even with the existing rules, developers could build affordable homes for EWS/LIG, but they are not doing so due to the complexity and rigidity of the existing registration and permit system. On the flip side, there are policies to provide for a higher Floor Area Ratio in the public and private lands, where there will be a provision of affordable housing construction. This will undoubtedly serve as an incentive for the private player to develop the land for commercial purposes and at the same time develop housing for the urban poor that are affordable and regulated and have proper municipal facilities. But such policies are only regionally administered. This year in the 2017 budget, affordable housing has been granted an infrastructure status, which, however, may provide an economic stimulus and easier financial credit for investors, making it a lucrative segment for investment. Infrastructure status will further simplify the approval process for affordable projects and create clear guidelines; the affordable housing segment may grow at a faster pace than the rest of the real estate sector and will be the key growth driver for the Indian mortgage finance market.

Pitfalls of "think centrally, act locally"

The Indian federal governing system has a hierarchical structure with a straight demarcation between the power and responsibilities of central and state governments as their electoral politics. This hierarchy ensures the centre's role in planning and funding for strategic infrastructure development, and the state governments are tasked with the implementation of various policies and programmes. With the advent of liberalization, state governments have enjoyed a higher degree of economic autonomy. However, due to the increasing regionalisation of politics, not all states have the willingness or capacity to deliver various policies. This arrangement has not been foolproof and there exist multitudes of pitfalls relating to technical issues (lack of data, inadequate skill sets of the personnel to execute projects in a sustainable manner), institutional capacity (budgetary constraints, lack of awareness/lack of opportunities for professional development) and the political will, collectively leading to the inadvertent creation of 'governance gaps'. The perception of ULBs merely as implementers in a centralised programme limits their ability to handle fiscal constraints or local challenges innovatively. The implementation of various housing programmes under decentralization involves a substantial reorganization of power and clearly defined roles for various local and state organisations so that each ULB can meet its target. Most importantly, the balance of power (roles and responsibilities) should be equal between the elected wing and the executive wing for an effective implementation of various projects/programmes and to transform ULBs into vibrant democratic units of self-government. At the local level, the conditions to create effective operation of the ULBs are still outstanding. Scholars argue that within the new approaches to delivery, public housing built by a contractor and chosen by a tender framework poses fresh challenges for already beleaguered ULBs. The consequence of trying to implement a more inclusive approach for in situ, multi-storey affordable housing with greater participation within the conventional public housing contractor model has been subject to greater fragmentation and lack of coordination within the ULBs (Kamath, 2012, p. 83).

New programmes so often laden with the notion of integrated and inclusive approach have forced the city and state governments to undertake a series of reforms as a condition for award of the grants. These include some of the complex reforms relating to certain policy measures (e.g. property tax and user charges for basic services). Under JNNURM for instance, the 'one-size-fits-all' approach of demanding that all states and ULBs achieve 23 reforms within seven years regardless of their stage of development, capacity and financial status, was over-ambitious (Thornton, 2011). The challenge of implementing both reforms and projects is beyond the capacity of many ULBs. The central-local mistrust in India is historical, manifested not only in the political environment but also in social and cultural realms. States closely allied with the governing party at the centre continue to fare much better in procuring central funds or echoing the central level policies compared to those that are ruled by the opposite parties. Moreover, there is always a tendency to misuse power, both at the state and central levels, for political mileage, thus hindering the long-term processes of development. Hence there is a huge diversity in terms of implementation of housing policies across the country,[10] with widely varying degrees of social development.

Participation, decentralisation and diversity

In India, Tiwari and Rao (2016) contend that while most of these programmes are well-intended in terms of their housing decentralisation objectives, they could not deliver much due to the lack of financial resources at the local level and excessive dependency on the central government for funds leading to sustenance of the top–down approach, with poor participation from the state governments, marginal inputs from the operational agencies and lack of public participation. Livengood and Kunte (2012) claim that NGO and CBO participation in BSUP projects, like those under Valmiki Ambedkar Awas Yojana, had no place in the decision-making process. The eligibility lists, house designs, specifications and terms and conditions are developed before NGOs and slum dwellers are invited to participate and bid on projects. Public and civic body participation in the decision-making of individual city development plans has remained a far cry despite the rhetoric of the Ministry of Housing and Urban Poverty Alleviation to encourage participation in the planning and decision-making phases. On the other hand, the lack of progress in capacity building among both ULBs and local governments to prepare and implement projects is striking. Most of the smaller ULBs do not have the capacity to prepare City Development Plans, but they endorsed it notionally so that projects could be submitted to the central government and funds could flow to cities (Sivaramakrishnan, 2011). The complex requirements set by RAY for technology-laden data collection and analysis before local plans and funding can be approved has been an exclusionary practice (Livengood & Kunte, 2012). This mirrors the technical requirements forced by JNNURM under BSUP that excluded most of the economically less advanced states. These inconsistencies are compounded by the considerable diversity of perspectives (such as *in situ* upgrading vs. slum redevelopment) and that multitudes of actors involved in the process are often at loggerheads with each other, considerably slowing down the process. The programmes such as JNNURM had set aside funds for capacity building, but very little has been drawn for this purpose. Roles of individual states were marginalised, and many had no appetite to drive ambitious programmes. Without active participation, the city master plans, developed by participating state governments, remain divorced from the urban planning process and lack connectivity. It is clear that decentralisation is central to providing a new context within which low-income housing in India is emerging. More precisely, decentralisation is taking place within the context of neoliberal national macroeconomic policy. While the concept is unproblematic, without adequate empowerment of local actors, the process breeds ambivalence and inconsistency.

Conclusion

This chapter traces shifts and continuities in recent positions on housing in India. The analysis has taken an empirical approach to examine the recent national housing programmes and used historical factors and international advocacies to explain their evolution, operation and implications. Judging by the scale of the operation, these programmes have evolved into projects of grand scale, with the dualistic approach of welfarist and economic development. As is evident, no single national programme in India, especially since the advent of neoliberalism, has been developed as a 'housing policy' on its own. This notion is embedded in the approach that sees subsidies directed to financing new, complete homes as economic 'assets' that can activate the construction and finance sector (Sengupta, Murtagh, D'Ottovanio & Pasternak, 2017) and could be used as collateral for formal loans and credits. The low-income housing delivery is thus seen as driven by economic policy rather than a social policy. The recent move towards 'Housing for all by 2022' by delivering up to 20 million units was launched to revive the private housing market and its productive capital stock. The government is now collaborating with private builders and developers under a public-private partnership (PPP) model to make 'Housing for All by 2022' a reality. This can be argued PAY is launched as a corrective measure given both JNNURM and RAY didn't have a strong 'housing' focus. The overarching pro-poor rhetoric of JNNURM is, however, riddled with inherent contradictions given that the decision to limit affordable housing content to just 35 per cent in majority schemes seriously undermines the objective, let alone the 'affordable branding', of these projects.

The overall lack of social 'quotient' of the programmes have been, inter alia, a result of governance gaps arising from the institutional arrangement for delivery. An organized housing market cannot exist without a set of institutional foundations that establish various programmes, oversee their implementation and ensure that the rights of low-income people are not compromised in the process. The appraisal of these programmes identifies a serious lack of resources and skill base of various ULBs. If ULBs are the institutional foundations, they need to be strengthened, regulated, and modernized as delivery vehicles. Moreover, for an effective implementation, it is important that there is one central-state-ULB framework to ensure better working together towards an integrated housing and social development goal.

Ultimately, state-led housing schemes should seek to put in place an effective finance mechanism in order to enable the urban poor. Both the World Bank and UN-Habitat have highlighted access to finance to be urgent, critical and necessary (UN-Habitat, 2016). However, the way subsidy is dispensed (through low-interest loans) randomly and as in government housing schemes largely ignores the real needs of the urban poor. There must be a fundamental break with the pragmatic calculations, which disfigure current housing finance thinking in India. Access to housing finance is restricted to middle- and upper-class people and is yet to penetrate the mass market owing to stringent eligibility and affordability criteria[11] (Sengupta, 2006; Smets, 1997 and transaction costs (Garg, 1998). As a result, the penetration of mortgage finance in India has remained miniscule – at just 8 per cent after decades of reform. The entry of the global agencies such as the World Bank into the housing finance sector has been successful to capture the 'prosperity' by targeting the upper middle-class households rather than to 'ending poverty'. While International agencies influence national policies, housing policy is a national responsibility and India is still lacking in developing a comprehensive housing policy framework that fully addresses these challenges.

Notes

1 Economically Weaker Sections (EWS) is a term used to refer to households with income below a certain threshold level. Though there may be other economic factors in deciding on the economic context of the household, income is the dominant criterion. There is no coherent single/unique definition for EWS in India and it is defined differently for different schemes run by the Government. States also have the flexibility to redefine the annual income criteria as per local context in consultation with the Centre. Under the new Housing Scheme – Prime Minister's Awaaz Yojna, EWS households are defined as households having an annual income of up to 300,000 INR (US$4,618). For Low Income Households (LIG), the income ceiling is between 3–6 lakh INR (US$4,618–9,236).

2 Inadequacies in access to formal finance have led to the growth of microfinance in India. Microfinance is defined as financial services such as savings accounts, insurance funds and credit provided to poor households (generally households who fall just above or just below the poverty line). The main features have been short-term working capital loans through voluntary or involuntary savings services. As a result,

microfinance has developed as an alternative to provide loans to poor people with the goal creating financial inclusion and equality. 'In India, it has reached over a 101 million households through 7.9 million Self Help Groups accounting for approximately a portfolio of 136,914 million INR' (Santosh et al., 2016).

3 See International Finance Corporation (2017).

4 Biswas, Kidokoro & Seta, (2017) observed that the term 'inclusive growth' was first introduced in India with the government's 11th Five-year Plan (2007–2012) and was updated in the 12th Five-year Plan on how to achieve more sustainable and inclusive growth.

5 *Bustees* are characterized by poor physical and environmental conditions (similar to their other variants–slums or squatter settlements) but are recognised as 'urban units' by the municipal authority having a 'legal tenure' but often having a complex and multi-tier ownership system. In Kolkata it is common to find *bustees* consisting of a three-tier tenancy system with the landowner in the first tier, the hut-owner who has the lease of the land (Thika Tenant) on the second, and the *bustee* dwellers to whom the huts have been rented on the third tier. These transactions and transfer of rights are regulated.

6 Urban Basic Services Scheme (1986), later renamed as Urban Basic Services for Poor in 1991); Nehru Rozgar Yojna's Scheme of Housing and Shelter Upgradation (1990) and the National Slum Development Programme (1996) were among those.

7 Income classification has been the benchmark for welfare distribution in India. In recent years, owing to the complexity of defining incomes of a very large population, the central government has been less rigid in setting out thresholds. States and UTs have the flexibility to redefine the annual income criteria as per local needs with the approval of Ministry. The Ministry of Housing and Urban Affairs, in its latest Report, has revised the income thresholds for the EWS and LIG upwards to households with an annual income up to 300,000 INR (US$4,500) and 600,000 INR (US$8,960), respectively (MoHUPA, 2015).

8 Under the JNNURM, the central government gives grants covering 50 per cent of the project cost for cities with populations between one million and four million. For cities with populations higher than four million, the central grant is 35 per cent of the project cost. The remaining funding comes from the state's kitty and the urban local bodies or parastatals. At present, there are 523 projects related to urban infrastructure development that are being implemented in 65 cities across the country.

9 *Crore* denotes 10 million.

10 States with poor standards of governance have shown ability over time to emulate the achievements of the leading states, lifting them from the ranks of the worst performing states (in housing provision) to being at par with those on the top. Smaller states such as Chattisgarh and Jharkhand have significantly leapt forward in the last 10 years in terms of per capita public investment in essential sectors such as housing. While poorer states such as Bihar and Uttar Pradesh still lag behind, they have in recent years enjoyed the political and economic focus with the BJP supporting governments in power.

11 Apart from viewing affordability as an income problem, it can also be seen as a housing market problem. Examining the formal housing finance in India, Smets's (1995, 1997) research illustrates that the formal housing finance is fundamentally flawed as a concept where households are expected to construct complete housing units in one go in the formal housing market. In practice, the urban poor tend to build their houses step by step. The sheer lack of concept and mechanism for 'incremental building and financing' in the formal housing vocabulary has in a way jeopardized the use and access of formal housing finance in the housing markets of developing countries.

References

Baindur, V. & Kamath, L. (2009). *Reengineering urban infrastructure: How the World Bank and Asian Development Bank shape urban infrastructure finance and governance in India.* New Delhi: Bank Information Centre.

Biswas, A., Kidokoro, T. & Seta, F. (2017). Analysis of Indian urban policies to identify their potential of achieving inclusive urban growth. *Urban Research and Practice, 10*(2), 198–227.

Castells, M., Goh, L. & Kwok, R. Y. W. (1990). *The Shek Kip Mei Syndrome: Economic development and housing policy in Hong Kong and Singapore*. London: Pion.

Choguill, C. L. (1997). Ten steps to sustainable urban infrastructure. *The Urban Age*, 5(2), 22–23.

Datta, K. & Jones, G. (Eds.) (1999). *Housing and finance in developing countries*. London: Routledge.

Economic and Political Weekly (EPW). (2017). *Monstrous Indian income inequality*, 52(40), 7 October 2017.

Garg, Y. K. (1998). *New directions in housing finance: India case study*. Paper prepared for the International Finance Corporation, presented at the Asian Housing Finance Workshop, Bali, Indonesia, 5–6 February 1998.

Government of India. (2011). *Indian census data*. Ministry of Home Affairs, Government of India.

International Finance Corporation. (2017). *IFC offers a wide variety of financial products for private sector projects in developing countries*. Available at www.ifc.org/wps/wcm/connect/corp_ext_content/ifc_external_corporate_site/home, accessed on 9 June 2017.

Jolly, R., Emmerij, L. & Weiss, T. G. (2009). *UN ideas that changed the world*. Bloomington: Indiana University Press.

Jones, G. A. & Ward, P. M. (1995). The World Bank's new urban management programs: Paradigm shift or policy continuity? *Habitat International, 18*(3), 33–51.

Kamath, L. (2012). New policy paradigms and actual practices in slum housing. *Economic and Political Weekly, 47*, 37–48.

KPMG. (2014). *Indian real estate opening doors*. Delhi: KPMG.

Livengood, A. & Kunte, K. (2012). Enabling participatory planning with GIS: A case study of settlement mapping in Cuttack, India. *Environment and Urbanization, 24*(1), 77–97.

Mahadevia, D. (2011). Branded and renewed? Policies, politics and processes of urban development in the reform era. *Economic and Political Weekly, 46*(31).

Marcuse, P. (2012). A Critical approach to solve the housing problems. In N. Brenner, P. Marcuse & M. Mayer (Eds.), *Cities for people, not for profit: Critical urban theory and the right to the city*. Routledge: Abington, pp. 215–230.

McKinsey Global Institute. (2016). *India's urban awakening: Inclusive cities, sustaining economic growth*. Delhi: MGI.

Ministry of Housing & Urban Poverty Alleviation, Government of India (MoHUPA). (2011). *Report of the working group on financing urban infrastructure: 12th five-year plan*. New Delhi: Steering Committee on Urban Development & Management, Government of India.

Ministry of Housing & Urban Poverty Alleviation, Government of India (MoHUPA). (2015). *Annual report 2015–2016*. New Delhi: Government of India.

Ministry of Housing & Urban Poverty Alleviation, Government of India (MoHUPA). (2016). *Pradhan Mantri Awas Yojana [Housing for all (urban) scheme guidelines]*. New Delhi: Government of India.

Patel, S. B. (2016). Assignment delivered, accountability nil: Housing for all by 2022. *Economic and Political Economy Weekly*, *51*(10), March 5, 38–42.

Rahman, M. M. (2001). Bastee eviction and housing rights: A case of Dhaka, Bangladesh. *Habitat International*, *25*(1), 49–67.

Ram, P. & Needham, B. (2016). The provision of affordable housing in India: Are commercial developers interested? *Habitat International*, 55, 100–108.

Rolnik, R. (2013). Late neoliberalism: The financialization of homeownership and housing rights. *International Journal of Urban and Regional Research*, *37*(3), 1058–1066.

Santosh, K., Subrahmanyam, S. E. V. & Reddy, N. (2016). Microfinance – A holistic approach towards financial inclusion. *Imperial Journal of Interdisciplinary Research (IJIR)*, *2*(9), 1130–1137.

Sengupta, U. (2006). Liberalisation and the privatization of public rental housing in Kolkata, *Cities*, *23*, 269–278.

Sengupta, U. (2013). Inclusive development? A state-led land development model in new town, Kolkata. *Environment and Planning* C, *31*(2), 357–376.

Sengupta, U. (2014). New frontiers and challenges. In J. Bredenoord, P. V. Lindert & P. Smets (Eds.), *Affordable housing in the urban global south: Seeking sustainable solutions*. London: Earthscan, pp. 137–153.

Sengupta, U. & Tipple, A. G. (2007). Performance of public sector housing in Kolkata in the post reform milieu. *Urban Studies*, *44*(10), 2009–2027.

Sengupta, U., Murtagh, B., D'Ottovanio, C. & Pasternak, S. (2017). Between enabling and provider approaches: Key shifts in the national housing policy in India and Brazil, *Environment and Planning* C. doi: 10.1177/2399654417725754.

Sivaramakrishnan, K. C. (2011). *The revisioning Indian cities: The urban renewal mission*. New Delhi: Sage Publications.

Smets, P. (1995). Poor and in need for a house in the city! The government is there to help, but who is really helping who? *Nagarlok*, *28*(4), 78–92.

Smets, P. (1997). Private housing finance in India: reaching down market? *Habitat International*, *21*(1), 1–15.

Thornton, G. (2011). Appraisal of Jawaharlal Nehru National Urban Renewal Mission (JnNURM). *Final Report, Volume I*. Delhi: JNNURM.

Tiwari, P. & Parikh, J. (2012). Global housing challenge: A case study of CO2 emissions in India. *Spandrel Journal*, 5, 96–104.

Tiwari, P. & Rao, J. (2016). Housing markets and housing policies in India. *ADBI Working Paper Series*, Manila: Asian Development Bank Institute.

UN-Habitat. (2006). *Enabling shelter strategies: Review of experience from two decades of implementation*. Nairobi: UN-Habitat.

UN-Habitat. (2016). *World cities report*. Nairobi: UN-Habitat.

World Bank. (1993). *Public/private partnerships in enabling shelter strategies*. Nairobi: World Bank.

World Bank. (2010). *Systems of cities: Harnessing the potential of urbanization for growth and poverty alleviation*. Washington, D.C.: World Bank.

Yap, K. S. (2015). The enabling strategy and its discontent: Low-income housing policies and practices in Asia. *Habitat International*, 54(3), 166–172.

10 Housing affordability problems of the middle-income groups in Dhaka, Bangladesh

A policy environment analysis

Sadeque Md Zaber Chowdhury

Introduction

Housing is considered one of the most important urban issues in both developed and developing economies. Every household should be able to afford a decent home without impairing their ability to afford other basic necessities. Enabling the moderate-income groups to afford an acceptable housing standard is a major policy concern for countries all over the world, and this has been a core issue in the housing policy discourse since the early 1990s (Bramley, 1993, 1994; Freeman, Chaplin & Whitehead, 1997; Katz, Brown, Turner, Cunningham & Sawyer, 2003; Linneman & Megbolugbe, 1992; Maclennan & Williams, 1990; Whitehead, 1991). The magnitude of the problem of affordable housing is in fact more acute in developing countries (UN-HABITAT, 2011). Cities in developing countries are growing very rapidly. Coping with the increasing pressure for housing and other public facilities is a great challenge for the governments of developing countries. Bangladesh is a case in point.

The problems of housing affordability of the middle- and lower-middle income groups of Dhaka, the capital of Bangladesh, are examined in this chapter. The concepts and definitions of housing affordability vary depending on the economic and social contexts of specific countries. However, irrespective of the context, housing affordability is influenced not only by the market conditions, but also by the prevailing policy environment, among other social and economic factors. The impact of the supply-side instruments of the policy environment, such as the regulatory regime, on the provision of affordable housing and housing affordability has been widely studied, mainly in the context of developed or richer developing countries, where strong regulatory and institutional frameworks exist. Little has been done in the context of developing countries with weak regulatory and institutional frameworks. Therefore, the theoretical impact of regulations on housing markets is not clearly understood in places with weak enforcement of regulations. This chapter is based on the author's pioneering unpublished study of this kind in the context of Bangladesh (Sadeque, 2013). The chapter investigates the problems of housing affordability for the middle- and lower-middle income groups of Dhaka, the capital city, as well as the prime metropolitan of Bangladesh, and the study identified the underlying supply-side causes of the policy environment. The housing policy environment in Dhaka is examined to find out the nature and direct causes of the housing affordability problems faced by the middle-income groups of Dhaka. As land and urban planning regulations, infrastructure regime, and institutional factors are important supply-side instruments of a housing policy environment, they constitute the core of the analysis.

The context

The middle-income groups: definition and significance

There is no official definition of income groups in Dhaka. In Bangladesh, the members of the middle-class belong to the learned professions and hold strong educational qualifications and skills (Sen, 2007; Siddiqui, Qadir, Alamgir & Huq, 1990). Within the economic and social context of Dhaka, the middle-income groups are defined by Sadeque (2013) as those whose monthly income ranges from approximately US$160 to US$1,220. The middle-income group has further been divided into lower-middle-, middle-middle- and upper-middle-income groups as presented in Table 10.1.

In Bangladesh, the middle-income groups started to become a dominant power propelling economic growth at the end of the colonial period, during 1947. So far, the middle- and lower-middle income groups of Dhaka have performed the core economic activities of the country. They are the most educated and skilled segment of the population; yet, surprisingly, they are the most neglected sector of the community. They suffer from inadequate, unaffordable housing and community facilities, among other problems. Their voices are usually not taken into consideration when development plans, big or small, are formulated and implemented (Rahman 1991; Sadeque, 2013). In spite of the importance of the middle-income group in shaping the country, the housing affordability problems of that group has not received proper attention either in academic research or in policy formulation. This chapter intends to fill the gap based on the author's former unpublished study (Sadeque, 2013).

Table 10.1 Different income groups in Dhaka

Income group	*Monthly income range (BDT)*
Marginal	<5,000
Low income	5,000–13,199
Lower-middle income	13,200–24,999
Middle-middle income	25,000–49,999
Upper-middle income	50,000–99,999
High income	100,000<

Source: Sadeque (2013).

Note
1 BDT= 0.012 USD (as of November 2017).

The estimated urban population in Bangladesh is 38 million; the concentration of urban population is greater in the large cities, with more than 55 per cent of the urban population in six metropolitan cities (BBS, 2011; United Nations, 2014). The urban areas of Bangladesh, especially the metropolitan areas, are particularly challenged by the huge shortage of housing, and consequently the severe lack of housing affordability. The explosive and uncontrolled growth of the urban population and urban areas in Bangladesh and the paucity of resources are exacerbating the housing affordability problems of middle- and lower-middle-income groups. The extent of the problem is greater in Dhaka, where people from almost all socio-economic backgrounds are facing this problem although the nature and magnitude of the problem are different. The urban destitute require rehabilitation, low-income families need low-cost flats or plots and the middle- and lower-middle-income groups complain about the cost of a plot or a decent flat being far beyond their means.

While acknowledging that the low-income group in Dhaka also suffers from severe housing affordability problems, this study argues that there is a great need to conduct research on the housing affordability problems of the middle and lower-middle groups to provide a comprehensive picture. Questions may be raised about whether solving the problems of middle-income groups is inimical to the interests of the poor. Most researchers do not agree. In contrast, a solution would benefit the poor and does not imply a lack of concern for the poor, as Birdsall argues. The Asian Development Bank (ADB, 2010), in its report on the rise of Asian middle-income group, quotes from Birdsall (2010, p. 160):

> in the advanced economies the poor have probably benefited from rule of law, legal protections, and in general the greater accountability of government that a large and politically independent middle class demand, and from the universal and adequately funded education, health and social insurance programs a middle class wants and finances through the tax system … A focus on the middle class does not exclude a focus on the poor but extends it, including on the grounds that growth that is good for the large majority of people in developing countries is more likely to be economically and politically sustainable, both for economic and political reasons.

In spite of acute housing affordability problems in Dhaka, 'affordable housing' has received minimal attention in the policy agenda. Ghafur (2004) and Rahman (2003) observed that state intervention in housing served the interests mainly of the influential social groups and consequently resulted in rising land values, unfairly planned urbanization, rent exploitation and land speculation. It is almost impossible for even a higher-middle-income household to buy a house from the formal market due to excessive prices. Even in the rental market, the housing rents are so high that middle-income families need to compromise their living standards (Hoek-Smit, 2002). The prevailing housing market characteristics are depicted in later parts of the chapter. The government organizations responsible for planning and housing have failed to understand the forces that influence and shape the housing affordability and do not have any goal that can drive them to initiate appropriate programs for ensuring an adequate supply of affordable housing.

Policy environment

As the housing sector is comprised of a complex network of actors ranging from builders and lenders, manufacturers and suppliers of building materials to land developers, real estate agencies, architects, engineers and government agencies, a interrelation among all those actors is necessary for an efficient market. However, this interrelationship is very fragile in Bangladesh, which is worsening the housing affordability problem there, especially in Dhaka (ADB, 2010; CPD, 2008; Sadeque, 2013; World Bank, 2007). The affordability problems are further exacerbated due to the absence of an integrated urban planning system, comprehensive policy on urbanization and urban poverty and the lack of a well-equipped agency to implement such a policy (Farzana, 2004; World Bank, 2007).

The instruments of a well-functioning housing and land market as outlined by Shlomo Angel (2000) in his concept of a 'housing policy environment' constitute the conceptual framework of this chapter. The concept can be considered as a modified elaboration of the World Bank's enabling strategies, which aim at improving the performance of housing sector as it has been developed based on the World Bank's housing policy paper published in 1993, when it significantly drew the attention of many scholars and housing experts (Jones & Ward, 1994, 1995; Jones & Datta, 2000; LaNier, Oman & Reeve, 1987; Malpezzi, 1994; Pugh, 1995).

Angel's idea of a 'housing policy environment' consists of a set of interventions in the housing sector by different government agencies to motivate and enable housing actions. Interventions by public agencies may be a two-edged sword as not all the interventions may sustain a well-functioning housing sector. On the one hand, appropriate interventions can help in achieving a well-functioning housing market while on the other hand, these interventions, if not made efficiently, may constrain housing actions. Therefore, depending on the extent and nature of public sector interventions, the housing policy environment may be enabling or non-enabling (Angel, 2000). An enabling housing policy environment largely affects the provision of affordable housing as presented in Figure 10.1.

Housing affordability problems in Dhaka

Housing price trends and home ownership affordability

There is no common pricing strategy among the developers in Dhaka. The pricing of apartment units depends largely on the respective developers' choice as well as on the size and location. Apartment prices vary significantly even in the same location depending on the developer's reputation. However, the

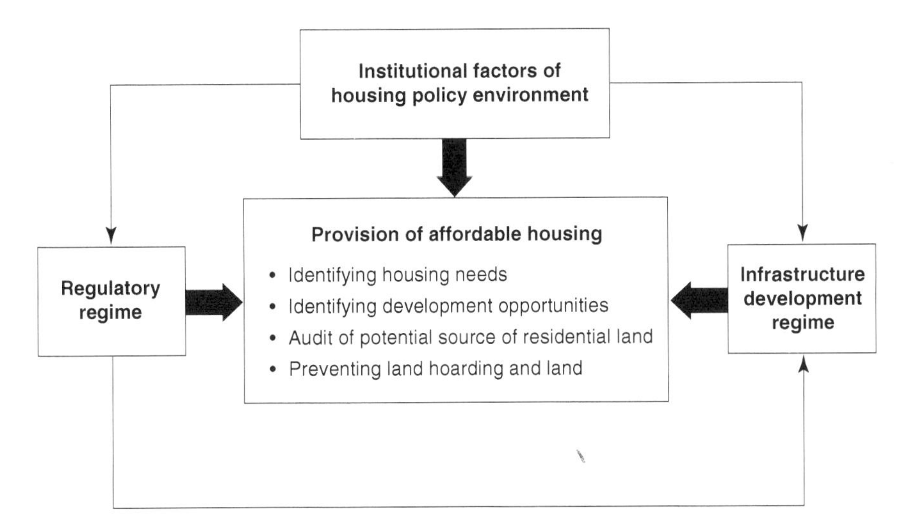

Figure 10.1 Thematic linkage between provisions of affordable housing and supply-side instruments of policy environment.

Source: Author.

median price per square foot for an average-sized apartment was found to be 3,600 BDT (US$44), while the price range was very wide, starting at 3,000 BDT (US$37) to as high as 14,000 BDT (US$170) per square foot; the median size of apartment units was 1,000 square feet. There are multifaceted explanations for this high price of apartment units in Dhaka, e.g., shortage in residential land supply, high price of construction materials, etc. (Sadeque, 2013).

Housing price experienced an increasing trend and the percentage increase in apartment price was exceptionally high (200 per cent) during the last decade (2000 to 2010). There was apparently quite a sharp rise in the price (115 per cent) during the years 2005 and 2010. Moreover, the increase in the price of housing was much higher than that of the nominal household income in Dhaka, as presented in Figure 10.2.

The housing price in designated high-income areas (as per price zoning) was much higher than the median price; however, the housing price even in the so-called 'middle income areas' was also increasing unprecedentedly (Sadeque, 2013).

The registration cost for an apartment in Dhaka was nearly 20 per cent of the apartment price, which was much higher than that in other countries. For example, in Hong Kong, the registration cost ranged from 1.8 per cent to a maximum of 6.8 per cent based on the value of the asset. In Malaysia, the registration fee ranged from 1.4 to 4 per cent based on the value of the property (Global Property Guide, 2012; HKSAR Inland Revenue Department, 2012).

In Dhaka, getting any readily available solid data for the purpose of calculating the house price-to-income ratio is very difficult. A study conducted by

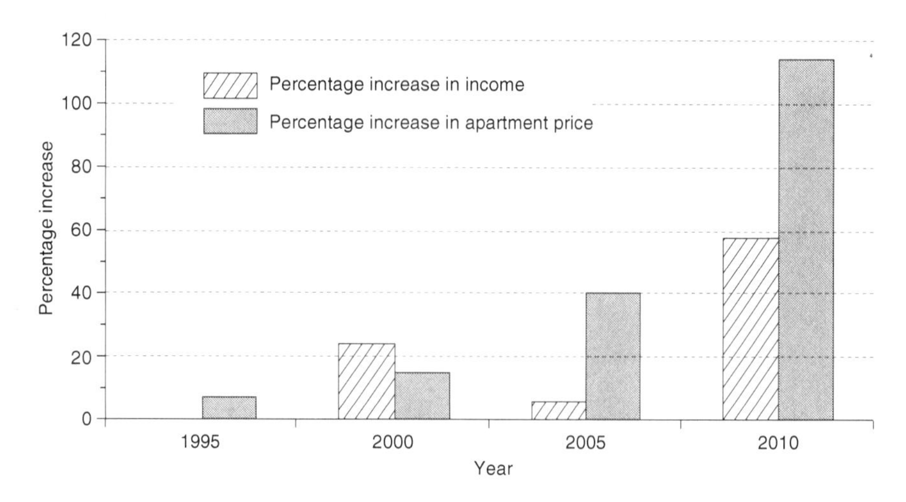

Figure 10.2 Percentage increase in housing price against increase in nominal household income.

Source: Sadeque (2013).

ADB (Hoek-Smit, 1998) shows that the house price-to-income ratio in Dhaka was 18.5 in 1993, and this is the only available data anywhere in the government documents. However, a more recent study by UN-HABITAT reveals that the ratio was 16.7, although the data source was not mentioned (UN-HABITAT, 2011). The study by Sadeque (2013) revealed that this ratio was surprisingly high and almost four times (18.8) the internationally acceptable ratio of 5.0 in 2013 (Figure 10.3). Moreover, this ratio in Dhaka was significantly higher than that of the major cities of India and Sri Lanka, both of which enjoy better economies than does Bangladesh, as shown in Figure 10.4 (Sukumar, 2001; UN-HABITAT, 2011).

Further, as calculated by Sadeque (2013), Table 10.2 represents the housing affordability using the residual-income approach.

Table 10.2 Housing affordability using residual-income-based approach

Income group	*Average monthly income (BDT)*	*Monthly housing cost*	*Average residual income*	*The amount required to afford a minimum standard of life*
Higher-middle	75,000	40% of income	45,000	20,000
Middle-middle	37,500	40% of income	22,500	
Lower-middle	19,100	40% of income	11,460	

Source: Sadeque (2013).

Note
1 BDT= 0.012 USD (as of November 2017).

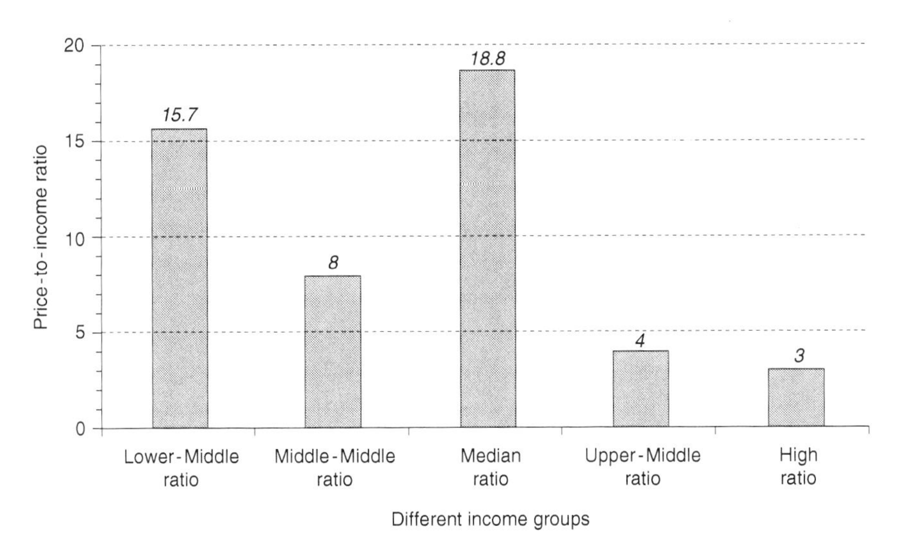

Figure 10.3 House price-to-income ratio in Dhaka.
Source: Sadeque (2013).

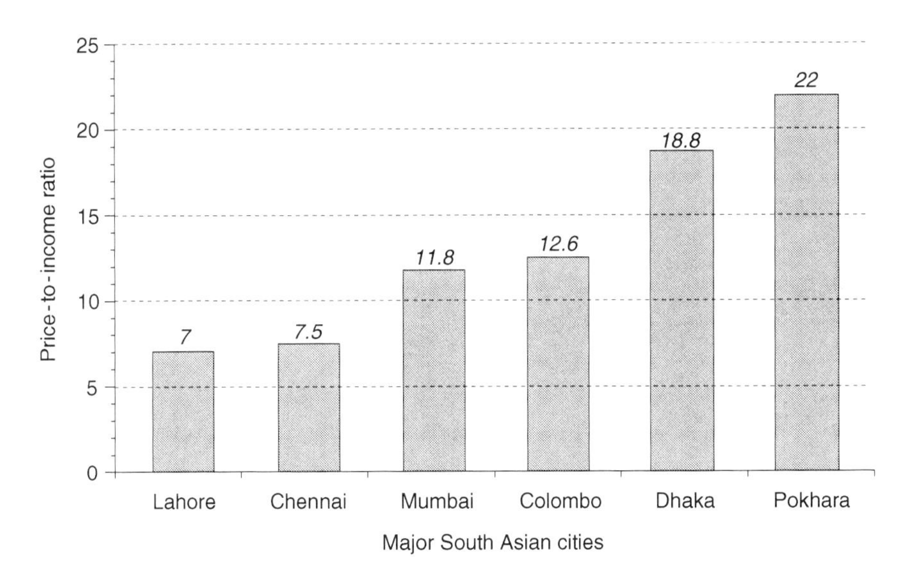

Figure 10.4 House price-to-income ratio in major South Asian cities during 2011.
Sources: Sadeque (2013), UN-HABITAT (2011).

Evidently, even the total monthly income of the lower-middle-income group was not adequate to maintain a minimum standard of living in the prevailing social and economic context of Dhaka then. The middle-middle-income group also suffered from affordability problems, although the magnitude of their problem was less than that of the lower-middle-income groups. Their residual income was slightly higher than the amount required to afford a minimum standard of living, but there were other big problems. People who belong to this income group had higher degrees and skills, and therefore it was understandably difficult for them to accept a minimum standard of living, including only the very basic needs without any recreation, good schools for their children, etc.

In Dhaka, a significant proportion of housing belongs to the informal sector and is of variable quality. Such informal housing may be affordable for the middle-income groups in terms of cost. However, this income group cannot live in informal settlements as the form and direction of the dominant social values are crucial parts of any housing context (Burke & Hulse, 2010). One of the important reasons why the middle-middle- and lower-middle-income families cannot live in low-cost informal settlements is the prevailing social context in Dhaka. The informal housing units are not of adequate standards, and a middle-middle- or lower-middle-income family does not intend to live there due to the prevailing social norms and status. While the middle-income households tend to rent housing units, the social norms invariably lead them to consider the social advantages and disadvantages of the intended location, such as the social status of the neighbors. Moreover, other factors such as the accessibility of community services and possible opportunities offered by the neighborhood are also considered (Stone, 1993). Thus, the middle-income group in Dhaka is caught in a dilemma. They can neither live a luxurious life like the high-income group nor live the way the lower-income group lives, because of the social norms and because of so-called social status. They cannot live in the slums as they need to send children to good schools and arrange and attend social gatherings. It seems that the middle-income group is trapped in the dilemma. Therefore, the middle-middle- and lower-middle-income households need to spend more than that they can afford for housing rent in the formal sector housing and consequently struggle to afford other daily necessities.

The median size of the apartment units in Dhaka was 1,000 square feet, and the median price per square foot was 3,600 BDT (US$44), which means that the median price of an apartment in Dhaka was 3.6 million BDT (US$45,000). The house price-to-income ratio in Dhaka has already revealed that such a price was not affordable for the middle- and lower-middle-income groups (price-to-income ratios of 8 and 15.7, respectively), and that such housing units were built exclusively for the upper-income group (Sadeque, 2013). Although the average size of apartment units in Dhaka was 1,000 square feet in the middle-income areas, the modal size of apartments was above 1,000 square feet, which indicates that the formal housing market was responding to the demands of only the higher – and a portion of the higher-middle-income groups (Sadeque, 2013). Paradoxically, it is evident that for the middle-income group, and especially for the middle-middle and lower-middle-income groups, there was a lack of housing units of affordable sizes, which implies that the housing market was unresponsive to their demands.

This indicator is used to identify the extent to which the formal sector is involved in building low-cost housing, and it helps in assessing whether the formal market is concentrating only on upper-end clients. Down-market penetration has been defined as the ratio between the price of the lowest-priced house produced and marketed by the formal private sector in significant quantities (not less than 2 per cent of annual housing production) and the median household income. This ratio is 19.5 in Dhaka and is slightly higher than the house price-to-income ratio, which implies that the formal sector has yet to reach below-median-income households.

Affordability problems in rental housing

As home ownership in Dhaka is a significantly expensive and difficult proposition, the middle-income class relies largely on rental housing. Rental affordability is usually measured in terms of the share of income that a household spends monthly for rent. In the context of a developed country, the standard threshold is 25–30 per cent of initial income spent on house rent, including utilities.

Households spending above a certain percentage of their monthly income for rent are classified as having a housing affordability problem. The median rent-to-income ratio in Dhaka was 69, which implies that a family with a monthly income equal to the median income (16,000 BDT or US$195) had to pay 69 per cent of its income as rent.[1] This ratio appears exceptionally high, and this

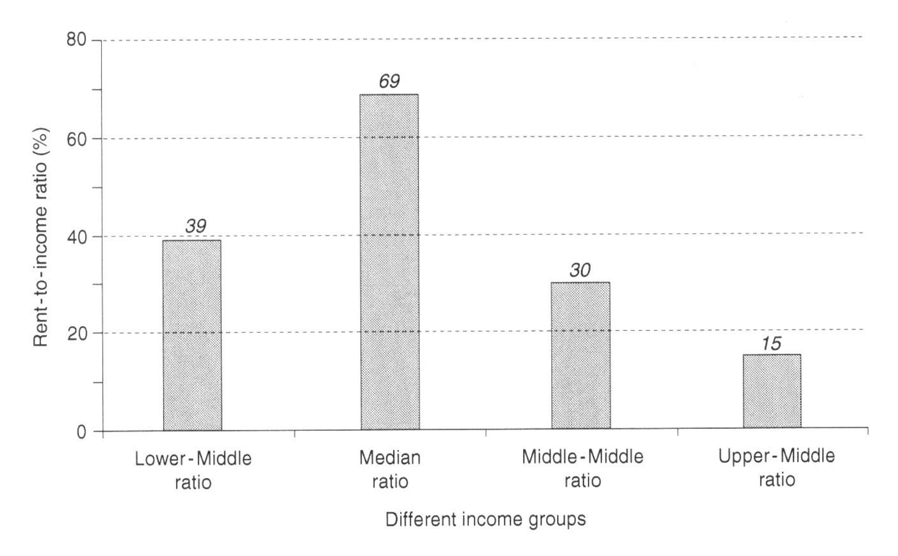

Figure 10.5 House rent-to-income ratio for different income groups in Dhaka.

Source: Sadeque (2013).

was because of the huge difference and variations in rent in the formal and informal sectors. Rent in the formal sector was at least double that in the informal sector, and in some instances even more; therefore, the median rent appears very high (Sadeque, 2013).

At 34:1, Pokhara in Nepal had one of the highest house rent-to-income ratios in Asia; however, in Dhaka this ratio was double (69:1) that in Pokhara and was the highest among all major South Asian cities, as presented in Figure 10.6. This ratio implies that Dhaka has the highest rent-to-income ratio in Asia.

Thus, most of the lower-middle and middle-middle income families in Dhaka are either 'severely cost-burdened' or 'cost-burdened' (Sadeque, 2013).

Land price in Dhaka

The percentage of land cost of total construction cost is important in influencing the affordable housing supply as it directly affects housing prices and rentals; in a well-functioning housing market, as the literature suggests, the acceptable proportion should be within the range of 25–30 per cent. Within the metropolitan area of Dhaka, the percentage of land cost to total construction cost was much higher than the standard value: on average it was 63 per cent, whereas the median value was 57 per cent of the total construction cost (Sadeque, 2013). Figure 10.7 depicts the prevailing scenario. The glaring affordability problems owe to the supply-side deficiencies in the policy environment.

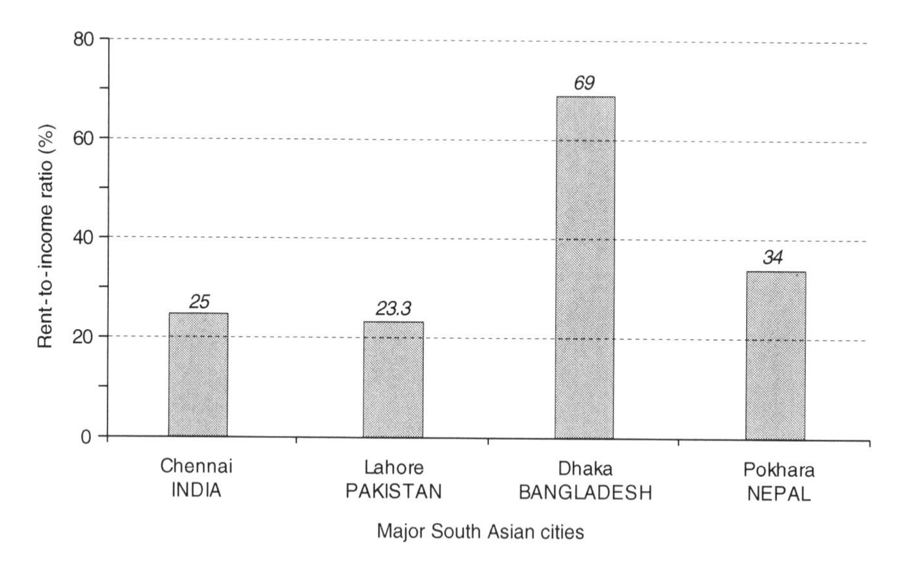

Figure 10.6 Housing rent-to-income ratio in Dhaka compared with other major South Asian cities.

Sources: Sadeque (2013), UN-HABITAT (2011).

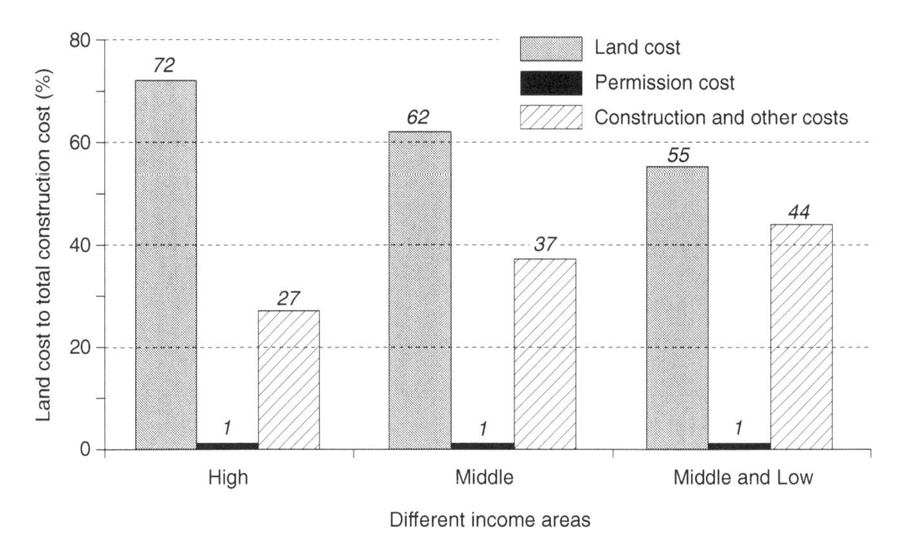

Figure 10.7 Percentage of land cost to total construction cost in different income areas in Dhaka.

Source: Sadeque (2013).

The policy environment of housing supply

The supply-side regulatory regime

The study by Sadeque (2013) revealed that regulatory regime was weak and fragmented, leading to a non-enabling housing market in Dhaka. The analysis of regulations showed that the prevailing regulations were not formulated in accordance with the local context to ensure an enabling housing and land market.

Dhaka has regulations related to land use, building construction, planning permission and residential land development. Mandatory standards were also set for residential land development and building construction. However, the regulations were prepared ignoring the provision of affordable housing, and by 2017, no comprehensive planning strategies have been proposed to ensure the supply of affordable housing for different income groups. No strict urban growth control mechanisms (i.e., green belt boundary, strict zoning) has existed in Dhaka. Therefore, their impact on land supply was insignificant. Despite the weak supply control, however, the land price was still very high, and the supply of residential land failed to meet the demand.

Moreover, while the mandatory standards for residential land development projects and construction of residential buildings (maximum saleable land, maximum gross population density, provision for community facilities and amenities, minimum road width, provision for water bodies, maximum ground coverage for building construction, uniform Floor Area Ratio (FAR), etc.) can enhance the living environment, they were critical barriers to enabling fuller utilization of the scarce developable land to optimize the supply of residential floor areas. The standards restricted the development of small-scale, low-cost land development projects that could provide affordable residential land for the middle-income groups. The Private Residential Land Project Development Regulations 2004 of Dhaka mandate the keeping of at least 30 per cent of the total land area to provide different amenities in any land development project in Dhaka. Therefore, only a maximum of 70 per cent of the land area of any such project is saleable. However, in practice, as discussed below, only 55–58 per cent of the land area is saleable as the developers are required to reserve land for primary and secondary roads, schools and other provisions; these areas are not included in the above-mentioned 30 per cent reserved land. These requirements are excessive, as Dowall & Ellis (2007) argue that the requirements for reserving more than 30 per cent of the total land area create perverse incentives for private developers and make it very difficult to produce affordable housing. This requirement of reserving a minimum of 30 per cent of the total project area and meeting other required standards applies to any land development project, irrespective of the target group and project size. There is no flexibility in standards or provision for bonuses or any favor in this regard for residential land development targeted to middle- or low-income groups, as practiced in other countries like Argentina (in the cities of Buenos Aires, Córdoba and Rosario) (Casazza et al., 2011) and Pakistan (in Karachi). This lack of flexibility works as a disincentive for developers to invest in small-scale development projects and also in projects for middle- or lower-middle-income populations.

The zoning regulations in Dhaka were ineffective, and their effect on residential land supply and building types was insignificant. Pervasive exclusionary zoning practices that increased the housing cost were replaced in an enabling environment with minimal nuisance prevention and ample opportunities to build small exclusive communities under the influence of the World Bank. Additionally, in a well-functioning housing and land market, zoning regulations must be enforceable, efficient and clear and should encourage mixed occupation of land among the competing commercial, productive and residential uses and among different income groups. Imposing unnecessary restrictions on builders and developers of residential land is unexpected. The builders and developers must be allowed to choose the most efficient combination of land, labor and capital for constructing any kind of housing development in any jurisdiction (Angel, 2000; Dowall & Ellis, 2007). However, the prevailing zoning regulations in Dhaka are ambiguous, with no restricting effect on residential land supply and building types. Moreover, the provision of inclusionary zoning, which would positively affect the supply of affordable housing, does not exist. The concept of 'inclusionary zoning' is one of the effective tools in the government armory that can help in providing an adequate supply of affordable housing. It is argued that 'inclusionary zoning' is probably the most commonly known mechanism for generating dedicated affordable housing stock

(Whitehead, 2007). There are different operational variations of this basic model. In one form, developers are required to contribute to the affordable housing stock in exchange for development rights or zoning variances. In the other form, developers must contribute in cash to the affordable housing fund. Some inclusionary zoning programs may be mandatory while others may provide incentives.

The Detailed Area Plan (DAP) of Dhaka has proposed seventeen types of land use zoning. The maximum amount of land areas is zoned for residential use. No direct restrictions on housing density, housing size or building height are imposed. However, there are indirect restrictions on development density through the implementation of the Floor Area Ratio (FAR), which governs the amount of gross floor area in buildings that may indirectly affect population density due to its interplay with flat size. The FAR in Dhaka varies depending on the plot sizes; however, the ratio is uniform for specific plot sizes irrespective of their locations and land values. The land value varies significantly between the central and fringe areas, and provision of service infrastructure is also limited and concentrated in the central areas. Therefore, a uniform FAR irrespective of location, land value and availability of infrastructure provision restricts the maximum utilization of residential land in the central areas and accordingly is contributing to high housing prices. This provision of a uniform FAR irrespective of location has also some other negative consequences such as imposing large costs on the city's economy as well as limiting the density of urban development. Furthermore, the demand for land across the city increases as more land is required for the same amount of floor space. Considering these consequences, the FAR in central parts of most large cities in the world is set at much higher levels and ranges from 5 to 15 (Bertaud, 2004; Dowall & Ellis, 2007).

However, in Dhaka, where the plot size in the central city area generally remains within 10 Katha (1 Katha equals 720 square feet), the maximum allowable FAR for such a plot is only 4.3. Lin, Mai & Wang (2004), by developing a model, demonstrated the effects of a uniform FAR irrespective of locations and concluded that as a consequence the uniform FAR encourages non-productive use of housing capital, raises the equilibrium of housing prices and lowers city growth. In contrast, an increasing FAR towards the city center boosts the productive use of housing capital and reduces housing prices, thus fostering the city's economic growth. Further, the restrictions on ground coverage coupled with the prescribed FAR do not allow high-density low-rise development, which affects the ability of individual families to build houses on small, or very small, plots of land, forcing them either to consume larger quantities of land at a higher cost or to live in flats. The supply of developable and serviced land is limited in Dhaka; therefore, maintaining an efficient intensity of land use in the context of competing demands on a limited supply of developable land is important. The relative distribution of population has major implications for the provision of public facilities such as transport, utilities and social infrastructure. Therefore, the control of residential density is important. Higher density residential developments should be located near major public transport interchanges and areas with adequate infrastructure facilities. That location will make it possible to capture development opportunities and at the same time to reduce expenditure on infrastructure provisions. Dhaka's zoning for residential use and the FAR did not consider the availability of transport and other infrastructure provisions while allocating land for residential use. The absence of density zoning and the presence of a uniform FAR hinder the reliable estimation of the population capacities of areas zoned for residential development, or conversely the estimation of the land area required to accommodate a given population. Reliable estimates are critical in ensuring adequate infrastructure and services to satisfy the needs of the future population and also to facilitate maximum utilization of scarce land resources.

Land-use planning provides the framework for future use and development of land and thus aims at securing the most efficient and effective use of land in the public interest while minimizing land use incompatibility. The effect of land-use regulations may vary from country to country depending on the context such as the nature of land ownership. Land-use regulations must be appropriate to the situation of a city if they are to function as a guiding and positive framework for residential land supply. Regulations certainly affect the supply of residential land and therefore can be used as a tool to achieve the goal of lower house prices overall by enabling more land to be provided. The experiences of the UK and Australia show that planning can influence the location of new affordable housing and thus help in providing affordable housing as required by setting necessary standards and alternatives. In cities in the UK, where land is freehold and mostly privately owned, strong planning controls on land-use exist, and thus there are opportunities for the government to make use of this planning tool to advance housing objectives, notably the supply of affordable housing (Dowall, 1992; Whitehead, 2007). In the Haryana province of India, for example, private developers are required to allocate 20 per cent of the total plots to the Economically Weaker Section (EWS) in order to obtain a license for development of any residential area. In the case of apartment housing, they are required to allocate 15 per cent of the total number of flats sanctioned in the scheme to EWS households at a fixed government rate. Similarly, the Maharashtra Housing and Area Development Authority allows a 20 per cent increase of the normally permissible floor space index (FSI) for schemes with at least 60 per cent of the tenements in the EWS category (UN-HABITAT, 2011). However, the land use regulations in Dhaka were also ineffective and encouraged land-hoarding as well as land value speculation. The existing land-use regulations in Dhaka in the form of the Detailed Area Plan (DAP) acknowledge that the middle- and low-income groups are facing severe affordability problems; however, the acknowledgement of the problems is not reflected in the designation of land for residential use. The impact of the regulations on land and housing supply and on housing affordability has been ignored in the formulation process of these regulations (Sadeque, 2013). The plan lacks a strong strategic orientation or clear vision to accommodate the projected population; rather, it proposes significant land areas for residential purposes without considering the provision of infrastructures and related consequences such as land speculation and land hoarding. The plan was supposed to provide guidelines for the provision of housing for different income groups, given the prevailing problems and future needs. No specific provision or planning tool is recommended in the plan for housing of middle-income groups.

The land development right is not restricted, and it belongs to the landowners. However, nothing is done to encourage the landowners or developers to construct affordable housing units although in countries like Australia and New Zealand, where development rights belong to the private landowners, various incentives (e.g., allowing higher densities for affordable housing provision) are awarded to the landowners or developers for the construction of affordable housing. Furthermore, the experience in the UK, especially in England, reveals that the separation of the development right and land ownership is the key to successfully using the land-use regulations to provide adequate affordable housing for different income groups. As in Bangladesh, land ownership in the UK belongs to the private citizens; yet unlike the case in Bangladesh, the land development rights belong to the government. Additionally, recent legislation in the UK (S106 of Town and Country Planning Act 1990) has allowed the mandate of affordable housing provision as a prerequisite for residential planning permission. S106 empowers the government by ensuring the government ownership of development rights; the power for the local government to reject any specific proposal; and the flexibility to negotiate on a site-by-site basis for contributions around an indicative, rather than prescriptive, development plan (Gurran & Whitehead, 2011; Whitehead, 2007).

No such provision is made in the land-use plan of Dhaka, and the problem of affordable housing has been totally ignored in the planning system generally and in the DAP specifically. Moreover, only 0.3 per cent (745 acres) of the available land area is designated for low-income housing. Worse still, such development cannot be guaranteed as the development right belongs to the landowners and 34 per cent of available land is designated for development of market-priced housing. Additionally, although significant land area has been designated for residential use, these lands cannot go onto the market due to the lack of infrastructure. This fact implies that the land-use regulations in Dhaka have formulated the blueprint, but in practice these regulations are not supported by government, as the necessary infrastructure has not been installed. The land-use regulations in Dhaka have clearly failed to make use of the control of development rights to address the housing needs of the city. Moreover, in Dhaka the necessity and importance of land-use regulation in ensuring affordable housing is not included in the planning policy framework. The designation of excessive land for residential use will encourage land speculation and land-hoarding.

The planning permission processes of both residential land development and building construction have been extremely time-consuming due to cumbersome bureaucratic planning procedures and the involvement of multifaceted government departments and agencies, which have imposed costs of delay.

For residential building construction, depending on the building size, these processes for obtaining planning permission have required 7 to 24 months. A residential land development project has required at least five years to get the planning permission. In other developed countries like the UK, it also takes a long time to obtain planning permission; however, this delay happens basically due to a more democratic practice in the form of public participation or due to negotiation on the development contribution (Ball, 2011). But, in Dhaka, there has been no such practice, and the extremely lengthy time requirement has existed due to regulatory and institutional complexities.

Further, the anti-speculative measures are also weak and in fact have encouraged speculation in Dhaka. High demand for land, when it outstrips the supply, is one of the important factors leading to land speculation. Several factors, both on the demand- and supply-sides, contribute to driving the land prices beyond the productive value of the land. There are several regulatory methods to reduce land speculation, and they usually include limitations on land ownership, excess land holding tax, capital gains taxes, legalizing land tenure (thereby increasing the availability of land), betterment tax and windfall tax as well as increased property taxes on vacant land. A justification for higher property taxation on vacant land is to encourage capital investments on land as it is often not being utilized to its full potential. In many cities of some developing Asian countries, there is additional tax on vacant urban land to curb urban land speculation. For example, in Penang, Malaysia, property tax is levied on vacant urban land at the same level as on other land: at six per cent of its annual value. In the municipalities and cities of the Philippines, there is a provision for additional vacant land tax, and the tax rate cannot exceed five per cent of the land value. In Dhaka, no tax is imposed on vacant land; therefore, one can easily enjoy a highly profitable venture by holding vacant land without paying any tax other than land development tax (which is very negligible compared to the land value). One has to pay 12.5 per cent of the annual rent of a land (with building) to the City Corporation as holding tax, and this does not apply to vacant land.

In a few cities of developing Asian countries, there is levy of property tax and windfall profit tax in order to reduce land speculation. The windfall profit (unearned increment) is the accretion of capital values of property that are not foreseen or in any degree due to the efforts made, intelligence or capital invested by the owner (UNESCAP, 1995). In other words, the profit that occurs unexpectedly as a consequence of some event not controlled by those who profit from it is a windfall profit. For example, the construction of a road by the government in an area will increase the price of land in that area; the landowners will enjoy a high profit even though they did not make any effort for it. In the Philippines, there is a provision for a windfall profit tax, and the intention was to curb land speculation and to induce investors to improve land as capital investments. The tax is especially high for short-term land ownership. In Dhaka, capital gains are considered ordinary income and taxed at the standard income tax rate. Moreover, the acquisition costs and related expenses are deducted from the sale proceeds for calculating taxable capital gains. There is also no regulation or ceiling on urban land ownership in Dhaka. However, in some South Asian countries, i.e., India and Pakistan, the regulatory framework includes laws (e.g., the Urban Land Ceiling Act in India) to restrict the amount of land that can be privately owned with the intention of reducing land speculation.

Moreover, relaxed planning regulations also work as an incentive for land speculation in Dhaka as the legitimacy of a building permission remains valid for three years and therefore gives owners the opportunity to hold the land for a long time without construction. They just need to complete the construction up to the plinth level within three years of getting the permission to ensure that the permission remains valid. They use this provision to hoard land for an indefinite time, which indirectly hinders the timely supply of housing. They obtain the building permission from the authority and hold it for a long time to maximize profit. This practice of builders is creating an artificial shortage of land and housing units.

In theory, it is argued that the relaxation of regulatory restrictions on land use and other related sectors will have positive impacts on the accessibility of suitable residential land for affordable housing development as was experienced in the case of Bangkok between 1980 and the mid-1990s (Keivani & Werna, 2001). On the contrary, the relaxation of regulatory restrictions, if implemented inefficiently, may encourage land speculation as was experienced in Chile between 1974 and 1982. The experience in Dhaka is that relaxed planning regulatory restrictions on land-use engender the 'use value' of land, making it much lower than its 'investment value' and adversely affecting the efficiency of the land market.

In developed countries, land and buildings are usually considered to provide a reliable but lower return in the long term than capital market investments. In contrast, in most of the developing countries, land and buildings are considered highly promising sectors for potential investment as there are limited alternative investments such as stocks, bonds and savings deposits (UNESCAP, 1995); Bangladesh is no exception in this regard. All these factors lead to serious land speculation problems in Dhaka and lead to an artificial imbalance in land supply.

Furthermore, the absence of a comprehensive approach and proactive government intervention led to the exclusion of moderate-income groups from entering the formal land and housing markets. The housing and land markets catered to the needs of the high-income groups only, as the builders and developers did not face any regulatory difficulties in constructing high-end housing or land units. The housing problems loom large, and the problem of affordable housing is acute in Bangladesh, especially in Dhaka. However, 'affordable housing' has not hitherto received attention in the policy agenda. The policy makers did not perceive housing as important for economic and social development; therefore, no vision has been set from the government (Sadeque, 2013). From the above discussion, it is clear that the regulatory regime is not conducive to the construction or development of housing or land for moderate-income groups in Dhaka.

The infrastructure development regime

The provision of infrastructure is an important element of urban planning that indirectly affects housing affordability through its direct effect on the provision of land supply; it is therefore an important tool for government in ensuring a regulated supply of residential land at the right time and the desired locations. From the perspective of the housing sector, urban infrastructure is comprised of basic physical networks, i.e., roads and walkways; water, sewerage and drainage; and power and telecommunication. Other than the above physical networks, the public-transport system, solid-waste disposal facilities, school, parks and playgrounds are other elements of the infrastructure regime that may affect affordable housing. Land produced through the improvement of infrastructure is an important source of supply. In this connection, the extension of the road network and public utilities aimed to link the peripheral lands to the metropolitan networks is vital. For an enabling housing policy environment, an efficient and equitable provision of residential infrastructure is essential.

However, the infrastructure development regime in Dhaka was weak and fragmented; it has ignored market signals and has failed to adequately cope with changing socio-economic conditions. The developers, both public and private, routinely face the problem of inadequate availability as well as significant delays in the provision of infrastructure such as roads, water, electricity, sewerage and gas supplied by public agencies. These factors increase the costs for developers. As in other developing countries, in Bangladesh, the government owns, operates and finances nearly all infrastructure with limited financial resources. In private residential land development projects, it is the developers who are required to provide on-site infrastructure within the project area. Yet they need not pay any sort of impact fees that could affect the price of land. Nevertheless, with the growing demands and competing claims on public resources from other priority areas, it has become increasingly difficult for the government of Bangladesh to sustain the open-ended financing of infrastructure in the country (ADB & JICA, 2004).

Public investment in infrastructure is not targeted to facilitate residential land supply and therefore creates bottlenecks in the supply of serviced residential land. Moreover, responsible agencies do not focus on servicing existing and underdeveloped urban land for efficient housing development. Serviced residential plots are therefore acquiring scarcity value, and the consequence is more expensive housing (Sadeque, 2013).

A number of government agencies have been engaged in providing infrastructure. However, no coordination among them existed. For example, the Capital Development Authority (RAJUK) is responsible for the overall planning of the city. The residential land supply is a critical issue, and it depends largely on the provision of infrastructure. Yet RAJUK does not have the authority to provide the infrastructure required for ensuring adequate supply of residential land. RAJUK can provide only transport infrastructure in a limited capacity; providing the other necessary residential infrastructures depends on several other agencies. A strong coordination among all the agencies is required to provide residential infrastructure focusing on an adequate supply of serviced urban land for efficient housing development. However, coordination among the agencies is rarely found, and each agency performs according to its respective strategies. Additionally, the transport infrastructure was not found efficient enough to reduce the travel time to the central city from the fringe areas, which in turn hinders the expansion of the housing market in those areas with relatively lower land prices than in the inner-city areas. The very high land development multiplier of 5.2 in Dhaka compared to the average value of 4 in developing countries suggests that the city has had a shortage of infrastructure provisions, creating serious constraints in serviced residential land supply (Angel, 2000; Sadeque, 2013). In other words, this figure implies that bringing urban services to raw land is especially difficult in Dhaka, and thus the process of transforming raw land to serviced land is constrained. This constraint affects the adequate supply of affordable land and thus housing (Angel, 2000; Arimah, 2000). A World Bank study also found that the infrastructure is inadequate for expansion of housing into fringe areas of Dhaka. Yet this constraint on development is not perceived by the concerned authorities, and there is no planning to provide adequate infrastructure to ensure the efficient supply of residential land. This lack is due to the inefficiency of the concerned authorities as well as the lack of attention to the housing and land matters in the policy framework (World Bank, 2010), which is still the case today.

The infrastructure expenditure per capita would be a good indicator of the government's investment priorities. The low value of this indicator generally suggests that the investment in this sector has not focused on servicing existing and underdeveloped urban land for the efficient and adequate supply of residential land. This lack usually leads to a high level of poor housing conditions, high appreciation of house prices, a high incidence of homelessness, a reduction in public sector housing provision and greater difficulty in attaining home ownership. However, because of the fragmented institutional set-up and administrative complexity in the infrastructure development sector, the required data could not be obtained. Due to the data constraints which still prevail today, the report of the Asian Development Bank on the Economic Growth and Poverty Reduction in Bangladesh (ADB & JICA, 2004) can be cited to provide an indication of the infrastructure development scene in Dhaka. The ADB report found that the infrastructure deficiencies act as a major hindrance on the country's development. The inadequate infrastructure coverage, poor management, and cost recovery and inefficiency of the infrastructure services (which are mostly publicly managed) have constrained the much-needed expansion of infrastructure services from meeting the growing needs of the economy. Clearly, the infrastructure development regime was not enabling in Dhaka.

The institutional factors

The institutional factors of the policy environment affect the efficiency of the performance of the housing sector as the effective formulation, implementation and review of rules; regulations depend largely on these factors. They influence the policies, systems and processes that relevant organizations use to legislate, plan and manage their activities efficiently and to effectively coordinate with others to fulfill their mandate (UNDP, 2010). Effective institutional factors can bring together all the related public agencies as well as the private sector and representatives of non-governmental organizations and community-based organizations, which in turn ensure that policies and programs benefit deprived groups and elicit their participation. The strength of institutions is vital in achieving national development goals.

The structure of governance has important implications for policy formulation and performance as the existence of a functional and equitable housing and land market depends largely on the efficient governance of this sector. Whether the mode of governance is participatory or elitist is important as it exerts different, though significant, influence on the performance of the organizations (Chiu, 2010). Investigating the role of stakeholders in the formulation and implementation of housing policies and programs as well as the prevailing line of authority in the institutional structure can provide a clear idea of the structure and mode of governance in this sector in Dhaka.

Although various public-sector agencies related to housing are under the same ministry, there is no clear line of authority, and all the agencies work according to their individual schedules. The multiplicity of institutions with overlapping responsibilities in the housing sector is a serious impediment to the efficient functioning of the institutions. The two most important authorities responsible for providing affordable housing in Dhaka, RAJUK and the NHA, have initiated several housing and residential land development projects without any coordination between them. No comprehensive program to solve the housing affordability problems of the middle-income groups was initiated, and each of the agencies performs according to its individual programs without paying any attention to the objectives of national housing policy. Moreover, the effectiveness of the public-sector initiatives in tackling the housing affordability problems of the middle-income groups has never been assessed by any of the agencies. The public sector's response to the affordable housing shortfall is insignificant. These initiatives are poorly targeted, in certain cases to households with incomes higher than the middle- and lower-middle-income groups, and at the same time are excessively directed at civil servants as the mode of governance is not participatory and the decision-making process is bureaucrat-led. Far from eliminating the housing problems, those public housing and land programs are exacerbating the housing affordability problems of the middle-income groups.

The human resource capabilities in the public-sector agencies in Dhaka are weak. The inappropriate appointments by government in the highest positions of relevant agencies coupled with the poor human resource capabilities hinder efficient formulation, implementation and timely review of rules and regulations regarding housing and land (Sadeque, 2013).

Conclusion

This chapter presents the complex relationship between housing affordability and the policy environment in the context of developing countries with weak institutional and regulatory structures. The housing policy environment is not enabling in such contexts. Pro-active government initiatives to strengthen the enabling functions of the regulatory regimes are required. Infrastructure investment should aim to facilitate the residential land supply. Further, institutional restructuring is essential. The policy-makers must understand the importance of housing and should conceptualize the role that the government can play in ensuring the provision of affordable housing. The housing sector plays an important role in the economic development of a country. It is a central force in sound economic development. Investing in housing holds similar importance to investing in transportation, power and communication. Housing investment contributes to the economy directly and indirectly, through various backward and forward linkages, and works as a tool for employment creation. Thus, expanding access to affordable housing has not only social or equity benefits but also clear economic benefits, where the housing market can contribute to the overall economic development of nations, cities and households. Therefore, a 'comprehensive approach', including the restructuring of the regulatory and infrastructure development regimes, and a strengthening of the institutional capacities, supported by a strong and consistent political will, is recommended to ensure an enabling housing policy environment.

Note

1 The data in this section were collected by the author solely for his PhD dissertation. Public data to calculate affordability are unavailable in the Bangladesh context and must be compiled based on researchers' own survey data.

References

Angel, S. (2000). *Housing policy matters: A global analysis*. Oxford: Oxford University Press.

Arimah, B. C. (2000). Housing-sector performance in global perspective: A cross-city investigation. *Urban Studies*, *37*(13), 2551–2579.

Asian Development Bank (ADB). (2010). *Key indicators for Asia and the Pacific*. Manila: Asian Development Bank.

Asian Development Bank (ADB) & Japan International Cooperation Agency (JICA). (2004). *Economic growth and poverty reduction in Bangladesh*. Dhaka: Asian Development Bank.

Ball, M. (2011). Planning delay and the responsiveness of English housing supply. *Urban Studies*, *48*, 349–362.

Bangladesh Bureau of Statistics (BBS). (2011). *Population & housing census 2011: preliminary results*. Dhaka: Bangladesh Bureau of Statistics.

Bertaud, A. (2004). The spatial organization of cities: Deliberate outcome or unforeseen consequence?, *Working Paper 2004–01*. Berkeley, CA: Institute of Urban and Regional Development, University of California at Berkeley.

Birdsall, N. (2010). The (indispensable) middle class in developing countries; or the rich and the rest, not the poor and the rest. In R. Kanbur & M. Spence (Eds.) *Equity and growth in a globalizing world*. Washington, D.C.: World Bank, pp. 157–187.

Bramley, G. (1993). The impact of land use planning and tax subsidies on the supply and price of housing in Britain. *Urban Studies*, *30*(1), 5–30.

Bramley, G. (1994). An affordability crisis in British housing: Dimensions, causes and policy impact. *Housing Studies*, 9(1), 103–124.

Burke, T. & Hulse, K. (2010). The institutional structure of housing and the sub-prime crisis: An Australian case study. *Housing Studies*, *25*(6), 821–838.

Casazza, J., Monkkonen, P., Reese, E. & Ronconi, L. (2011). *Regulations and land markets in the peripheries of three Argentine cities: Buenos Aires, Córdoba and Rosario*. Centro de Implementación de Políticas Públicas para la Equidad y el Crecimiento (CIPPEC), Buenos Aires.

Center for Policy Dialogue (CPD). (2008). *Strengthening the role of private sector housing in Bangladesh economy: The policy challenges*. Dhaka: Center for Policy Dialogue.

Chiu, R. L. H. (2010). The transferability of Hong Kong's public housing policy. *International Journal of Housing Policy*, *10*(3), 301–323.

Communist Party of Vietnam. (1991). *Strategy for socio-economic stabilization and development to 2000*. Hanoi: Party Congress VII, June 1991.

Dowall, D. E. (1992). Benefits of minimal land-use regulations in developing countries. *Cato Journal*, *12*, 431–444.

Dowall, D. E. & Ellis, P. (2007). *Urban land and housing markets in the Punjab, Pakistan*. Berkeley, CA: Institute of Urban and Regional Development, University of California at Berkeley.

Farzana, F. (2004). *Shortages of middle income owner occupied housing in Dhaka: failures of government or market?* Unpublished Master's thesis submitted to National University of Singapore.

Freeman, A., Chaplin, R. & Whitehead, C. (1997). Rental affordability: A review of international literature, *Discussion Paper* 88. Cambridge: Property Research Unit, Department of Land Economy, University of Cambridge.

Ghafur, S. (2004). Home for human development: Policy implications for homelessness in Bangladesh. *International Development Planning Review*, 26, 261–286.

Global Property Guide. (2012). *Buying Property*. Bristol: Global Property Guide.

Gurran, N. & Whitehead, C. (2011). Planning and affordable housing in Australia and the UK: A comparative perspective. *Housing Studies*, *26*(7–8), 1193–1214.

Hoek-Smit, M. C. (1998). *Housing finance in Bangladesh: Improving access to housing finance by middle and lower income groups*. Dhaka: Ministry of Local Government, Rural Development and Co-operatives.

Hoek-Smit, M. C. (2002). *Implementing Indonesia's new housing policy: The way forward*. Kimpraswil: Government of Indonesia and the World Bank.

Hong Kong Special Administrative Region (HKSAR) Inland Revenue Department. (2012). *Stamp duty rates*. Hong Kong: Inland Revenue Department Stamp Office.

Jones, G. A. & Datta, K. (2000). Enabling market to work? Housing policy in the 'new' South Africa. *International Planning Studies*, *5*(3), 393–416.

Jones, G. A. & Ward, P. M. (1994). The World Bank's 'new' urban management programme: Paradigm shift or policy continuity? *Habitat International*, *18*(3), 33–51.

Jones, G. A. & Ward, P. M. (1995). The blind men and the elephant: A critic's reply. *Habitat International*, *19*(1), 61–72.

Katz, B., Brown, K. D., Turner, M. A., Cunningham, M. & Sawyer, N. (2003). Rethinking local affordable housing strategies: Lessons from 70 years of policy and practice, *Report*. Washington, D.C.: The Brookings Institution.

Keivani, R. & Werna, E. (2001). Modes of housing provision in developing countries. *Progress in Planning*, *55*(2), 65–118.

LaNier, R., Oman, C. A. & Reeve, S. (1987). *Encouraging private initiative*. The Office. Washington, DC: Technical Support Services, Inc.

Lin, C. C., Mai, C. C. & Wang, P. (2004). Urban land policy and housing in an endogenously growing monocentric city. *Regional Science and Urban Economics*, *34*(3), 241–261.

Linneman, P. D. & Megbolugbe, I. F. (1992). Housing affordability: Myth or reality? *Urban Studies*, *29*(3/4), 369–392.

Maclennan, D. & Williams, R. (Eds.) (1990). *Affordable Housing in Britain and America*. York: Joseph Rowntree Foundation.

Malpezzi, S. (1994). "Getting the incentives right:" A reply to Robert-Jan Baken and Jan Van Der Linden, *CULER Working Papers*. Madison, WI: Center for Urban Land Economic Research, University of Wisconsin.

Pugh, C. (1995). Urbanization in developing countries: An overview of the economic and policy issues in the 1990s. *Cities*, *12*(6), 381–398.

Rahman, M. M. (1991). *Urban lower-middle- and middle-income housing in Dhaka, Bangladesh: An investigation into affordability and options*. Doctoral thesis submitted to University of Nottingham.

Rahman, M. Z. (2003). *Urban policy in Bangladesh: The state, inequality and housing crises in Dhaka City*. Master's thesis submitted to University of Calgary.

Sadeque, C. M. Z. (2013). *The Housing affordability problems of the middle-income groups in Dhaka: A policy environment analysis*. Doctoral thesis submitted to the University of Hong Kong.

Sen, R. (2007). *The changing middle class of Dhaka City and its impact on Bangladesh society*. Dhaka: The Asiatic Society of Bangladesh.

Siddiqui, K., Qadir, S. R., Alamgir, S. & Huq, S. (1990). *Social formation in Dhaka City*. Dhaka: The University Press Limited (UPL).

Stone, M. E. (1993). *Shelter poverty: New ideas on housing affordability*. Philadelphia: Temple University Press.

Sukumar, G. (2001). Institutional potential of housing cooperatives for low-income households: The case of India. *Habitat International*, *25*, 147–174.

UN Economic and Social Commission for Asia and the Pacific (UNESCAP). (1995). *Municipal land management in Asia: A comparative study*. Bangkok: UNESCAP.

UNDP (2010). *Capactiy development: Measuring capacity*. New York: United Nations Development Program.

UN-HABITAT. (2011). *Affordable land and housing in Asia*. Nairobi: United Nations Human Settlements Programme.

United Nations. (2014). *World urbanization prospects*. New York: Department of Economic and Social Affairs, United Nations.

Whitehead, C. M. E. (1991). From need to affordability: An analysis of UK housing objectives. *Urban Studies*, *28*(6), 871–887.

Whitehead, C. M. E. (2007). Planning policies and affordable housing: England as a successful case study? *Housing Studies*, *22*(1), 25–44.

World Bank. (2007). Bangladesh-Dhaka: Improving living conditions for the urban poor, *Bangladesh Development Series*. Washington, D.C.: World Bank.

World Bank. (2010). *World Bank data: Bangladesh*. Washington, D.C.: World Bank.

11 Vietnam's post-reform housing policies

Social rhetoric, market imperatives and informality

Hoai Anh Tran and Ngai Ming Yip

Introduction

Vietnam has undergone rapid transformation from a socialist to a market system since the economic reform (*Doi Moi*), and housing policy shows one of the most comprehensive transformations to a market orientation. The creation of an extensive legal framework to boost the private sector's housing development has brought about both a remarkable growth of the housing stock and greatly improved housing quality. Yet this legal framework also leads to the widening of housing inequality (Gough & Tran, 2009). While good quality commodity housing has been developed by the private sector, the majority of units are out of reach for the average income earners. The supply of affordable housing by the state has been severely lagging, as practical alternatives for the urbanites with less means are limited. Furthermore, the formal housing sector (in both the private and state sectors), which is regulated by institutionalised frameworks, was able to produce only a small share of new urban housing despite being heavily supported by state policies. The bulk of the urban housing supply (75 per cent) was in fact accomplished by incremental housing activities fostered by individuals, households and small entrepreneurs outside the formal system.

In this respect, Vietnam's housing system provides a complex and interesting case for policy analysis. On the one hand, the transition from centrally planned housing to a market orientation helps to highlight, and draws the focus to, the relationship between the state and the market in housing policies. The co-existence of a huge informal housing sector alongside the formal housing sector adds the dimension of informality in the state-society relation, illuminating the role of the "society" in the transformation process.

This chapter attempts to provide a critical analysis of housing policies in Vietnam with a focus on the relationship between the state and the market and special attention to policies of housing provision for vulnerable groups and the social consequences of such policies. The chapter highlights the contradiction and complexity of a dual system of formal and informal housing in which the socialist goals of a political legacy intertwine with market imperatives introduced by the neoliberal doctrines. In sharp contrast to other developed or transitional economies (e.g. the neighbouring China), housing development in

Vietnam was both state-led and people-led with the seemingly undisciplined informal sector being the anchor. The chapter is based on analysis of policy documents, housing statistics and international donor reports, as well as media coverage on housing issues, with a focus on affordable housing.

The shaping of housing policy in Vietnam

To understand how housing policy is formulated in the reform and post-reform era of Vietnam, one must take note of the underpinning forces that shape the economic and social development of the country as a whole. Three factors are of paramount importance: neoliberalism in urban and housing policy, socialist rhetoric that creates path-dependency and the supposedly weak capacity of the Vietnamese state in implementing the reform measures.

The first and foremost force is neoliberalism. In the 1970s, neoliberalist policy appeared to be the holy grail for economic and social development and further proliferated with World Bank programmes and IMF loans to less-developed countries. It is thus not surprising that when economic reform began in China in the early 1980 and Vietnam in the mid-1980s, neoliberal economic and social policies seemed like the epitome of capitalism and were quickly being adopted. This step led to the comment of Harvey (2005) on China that "[she has] definitely moved towards neoliberalization" (p. 151). Yet more critical analysis of the process of market liberalization in China led Nonini (2008) to conclude that at best a weak form of neoliberalism exists within a limited proportion of Chinese citizens. The case in Vietnam may not be the same; although China was a role model for Vietnam at the beginning of the economic reform, Vietnam's trajectories of reform are very different from that of China.

Unlike China, which benefited from foreign direct investment from overseas Chinese in Hong Kong and Taiwan, Vietnam has heavily relied on foreign direct investment and foreign aid from countries in Northern and Western Europe as well as Asian countries like Japan and Korea (Masina, 2012). It is not surprising to find "international organizations have long pressed for broader neoliberal restructuring as a means to 'civilize', 'modernize' and privatize the landscape through minimizing state intrusion in the market" (Schwenkel & Leshkowich, 2012, p. 388). It is also the alignment of an international push with self-interests of officials that reinforces the neoliberal momentum (Harms, 2012).

The second factor is the socialist rhetoric and legacy of the socialist system. After the economic reform, Vietnam is a de facto capitalist economy that has been increasingly incorporated into the global economic order. Yet politically, it remains a totalitarian one-party socialist regime, at least constitutionally. Socialist doctrines such as the party's absolute rule, and the state's dominance of the economy, still form the overarching guiding principles of state policy. The state holds on to its economic and fiscal power while promoting a "multi-sector commodity economy" (London, 2009). Holding high the socialist agenda of pursuing economic growth alongside social equity (Communist Party of Vietnam, 1991), the state invested extensively in social development programs such as poverty reduction and development of a system of social security. However, true to the socialist idea of a dominant state sector, social security policies take care only of people in the formal sector and thus exclude vulnerable groups such as migrants, ethnic minorities, the elderly with no access to incomes, etc.

The third factor is what may be considered as a weak state capacity. With the unchecked political power of the one-party authoritarian state, it appears that Vietnam should have no difficulties in carrying out the reform given the determination of the party-state for change. In fact, during 1992 to 1999 alone, there were 120 new laws with thousands of implementation regulations and decrees being issued to push the economic reform forward. However, most such regulations were unable to turn from "law on paper" to "law in reality", partly because of the resistance from local officials or residents but largely owing to the hastiness of law-making that resulted in confusion and even contradiction among different legislation (Yip & Tran, 2008). To rectify such discrepancies, more discretion is allowed for local officials in implementing such regulations (Quinn, 2002). Yet this allowance often leads to the abuse of power and increases the opportunities for corruption.

The administrative capacity for carrying out policy and the political capacity for mediating conflicts of interest in policy implementation could also be seen as weak. For instance, in handling illegal building extensions, instead of enforcing the law, local officials incline to use disorganised dialogical means in order to avoid direct conflicts with residents, which often results in the extensions being tolerated (Koh, 2006). Equally ineffective is the political capacity of local officials in dealing with power conflicts. A multiple command chain and interference from upper authorities makes Vietnam appear less efficient in carrying out planned developments (Yip & Tran, 2008).

A consequence of the weak state capacity is the prevalence of the "popular sector impulses" (McGee, 2009) held by many economically disadvantaged citizens who gained autonomy after the economic liberation (Higgs, 2003, p. 87). This situation stimulates the informal economy and creates "self-organising cities" (Geertman, 2007) which is manifested by the "footpath" economy of street vendors roaming pavements of the city (Higgs, 2003), self-built "popular" housing and bulging extensions in residential buildings (Koh, 2006).

Yet these seemingly dominant factors may not withstand scrutiny. Gainsborough (2010) argues that while many aspects of reform are influenced by market imperatives, the Vietnamese state has maintained a significant degree of continuity in ideas and practices from the era of central planning. The Communist Party still considers itself to conform to socialist ideals (London, 2009). Subsequent sections will illustrate how such seemingly neoliberal policies interact with the prevailing social and economic environment as well as with the socialist legacy in unique trajectories deviating from what the neoliberal policies intend to achieve.

Likewise, the weak state thesis is also being challenged. For instance, the absence of a state regulatory authority does not necessarily mean the absence of state power, and uncertainty can be a deliberate instrument of state rule (Gainsborough, 2009, 2010). The state is not a single entity (Painter, 2005), and the working of a fragmented set of institutions that appear to move in different directions can be powerful in forming outcomes "larger than the sum of [their] part[s]" (Gainsborough, 2010, p. 183). Empirical examples in the next section will outline how an eloquent informal sector, which had been perceived as the undesirable product of a weak state capacity, over-shadows the formal state sector in producing real housing outcomes.

Housing policies in transition

The socialist housing system: stratification and welfare for the deserved

Housing was provided as a social benefit by the state in the pre-reform era, with the state monopolising both the production and distribution of housing in the urban areas. In this de-commodified housing system, not only was rent set very low, but tenancy secured life-long occupancy and could even be passed to children. However, state housing then did not bear any "social" meaning as in counterparts in the West but was allocated with regard to employment (seniority or position in the hierarchy) and political merit (whether "valuable" to the party state) (Tran & Dalholm, 2005). This situation would inevitably lead to stratification within the same work unit, which was the basic unit of housing allocation. In fact, diversity among different work units was even more apparent with stronger work units (powerful ministries, local institutions with political power) able to provide better housing than weaker work units (such as schools, small factories and enterprises) (Tran & Dalholm, 2005).

In spite of the government's effort in meeting the housing demand, the war and the weak economy in the aftermath of the war made housing supply lag far behind demand. By 1989, when *Doi Moi* had just started, the housing shortage was acute (Evertsz, 2000). Overcrowding was also common, with an average living space per person in Hanoi of only 5.8 square metres. The use of the housing stock was over-stretched so that apartments were extended or subdivided without permissions (Figure 11.1) and illegal subletting and unauthorised transfer of state apartments were widespread (Geertman, 2007; Tran & Dalholm, 2005). As public housing was distributed exclusively to state employees, the vast majority of the urban households had to satisfy their housing needs on their own. Such housing, developed by individuals, was officially only referred to as "temporary housing" outside the formal sector. Hence its significance has been largely underestimated for decades.

Doi Moi *and the growth of popular housing*

Doi Moi, the economic reform, saw the transformation of housing from a state-allocated to a market-distributed system with a large number of laws, decrees and resolutions being enacted in creating an efficient market system to boost housing

Figure 11.1 Illegal extension of public housing.
Photo: Hoai Anh Tran.

production (Yip & Tran, 2008). One of the first measures of housing reform was to acknowledge residents' "land use right" by lifting the ban on self-building. Previous renovation and construction activities without official endorsement as well as informal transactions were legalized as a means of incorporating already existing housing built by people into the formal regulatory framework (UN Habitat, 2014).

The reform measures triggered a flood of self-built housing in all forms and shapes – termed popular housing – filling up empty lots in Vietnamese cities (Evertsz, 2000) or on sites of demolished old houses. Self-organisation was the main feature of popular housing (Geertman, 2007): they were financed and built by and through the initiatives of individual households, outside the official framework but of high quality and by no means slums (Figure 11.2).

In the formal sector, a variety of new initiatives was promulgated to boost housing production. Such initiatives include work-unit housing, the "state and people work together" scheme and site and services schemes. Yet only a few such official housing plans were actually constructed in the 1990s, and none of them were sustainable (Geertman, 2007). By contrast, it was popular housing produced through self-built activities that made the greatest contribution to the urban housing stock in the 1990s, on average 70 per cent of the new housing stock in Hanoi between 1995 and 2000 (Geertman, 2007).

Figure 11.2 Popular housing made up 70 per cent of new housing built between 1995 and 2000.

Photo: Hoai Anh Tran.

Large-scale development and commodification of housing

By the end of the 1990s, housing policies took a clear turn towards large-scale corporate-led development. This was partly attributed to the acknowledgement of the inefficient land use of the popular housing sector and the inadequate urban infrastructure that had been developed to support incremental small-scale development. Even more importantly, on the other hand, it was the concern over the "disorderly", "uncontrolled" state of small-scale development (Thanh uy Ha Noi, 1998, p. 3) that did not fit into the national goals of modernisation and integration into the global economy. In addition, large-scale corporate-led development was also a measure for the government to regain control over urban development (Tran, 2015).

Under the new directive, urban housing developments were to be master-planned and large-scale and needed to comply with the detailed plan approved by the central authority. Governmental resolutions and decrees were formulated to guide this corporate-led housing production with a specific decree targeting "planned, synchronous urban areas with technical infrastructure, social infrastructure, residential areas and other services" (Government of Vietnam, 2006, p. 1). The spatial scale should be at least 50 hectares, and only in exceptional cases 20 hectares (ibid., 2006). These developments (Figure 11.3) were

Figure 11.3 Corporate housing in new urban areas: Ciputra Hanoi International City (model).

Photo: Hoai Anh Tran.

encouraged by a range of legal and financial supports as incentives. Such measures have been successful in gearing the development of the corporate real estate sector. In the pursuit of quick and high profit, high-end housing targeting middle-class households was promoted (Tran & Yip, forthcoming). This targeting makes the majority of such new housing developments out of the reach of the majority of urban residents. Despite the policy that requires 20 per cent of the newly developed residential land (or 30 per cent of the new housing) to be returned to the local government as public housing (Hanoi People's Committee, 2001), the implementation has been sluggish. In fact, much of the public housing produced by such schemes was being used for relocation and thus did not result in much addition to the public housing stock.

With the termination of public rental housing production, low-income people have no access to housing. The state's involvement in housing support for the low-income and the poor is restricted to a limited number of small-scale for-sale projects targeting specific priority groups such as people who contributed to the revolution, war veterans and a small number of very poor households (UN Habitat, 2014). In fact, having a formal income is set as a condition of application for mortgages to purchasing developer-produced housing. This criterion excludes the majority of the urban poor (68 per cent of the labour force), who do not have records of formal income (World Bank, 2015).

On the whole, the housing reform in Vietnam has been successful in boosting general housing production and set up a thriving housing market for housing exchange. The average floor area per person increased from 9.7 square metres in 1999 to 16.7 square metres in 2009 (UN Habitat, 2014). Yet housing inequality has also been exacerbated. Despite the investment of substantial state resources into housing production by the corporate sector, this sector accounts for only a modest level of 15 per cent of the housing stock (UN Habitat, 2014). It is the continuously expanded informal sector,[1] often without formal authorization, that has contributed the bulk of new housing production, 75 per cent of the urban housing stock by 2015 (World Bank, 2015) (Figure 11.4). The informal sector was also the main provider of affordable housing for the urban poor (World Bank, 2015, p. 19).

Hence a dual housing system exists in Vietnam. There is a formal sector that has received most state support but has performed poorly in terms of output and affordability, yet the majority of new housing has been produced in a vast informal sector without the support of the state. The inclination towards large-scale development was a move by the state to re-establish control and improve efficiency in the formal housing sector to realise the state's vision of modernization and integration with the global economy. The significance of the informal sector is beginning to be acknowledged. A new housing law enacted in 2015 permits self-building and renovation by individuals with financial and policy support similar to that of the corporate sector. This support includes low interest loans and subsidies as well as policy support of exemption or reduction of land use fees.

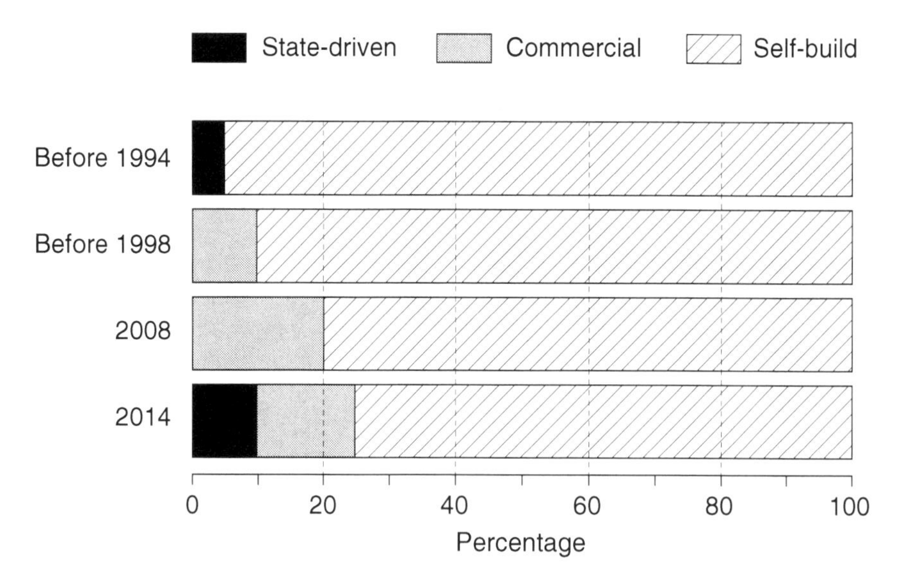

Figure 11.4 Change in housing production over time.

Source: World Bank (2015, p. 21).

Social housing: the path-dependent development of a concept

From the early 2000s, the government began to be concerned about the lack of adequate housing alternatives for low-income households after scrapping public housing construction for more than a decade. Social housing was reintroduced. The first move was to set up "priority" housing schemes (Government of Vietnam, 2001, p. 1) with incentives to encourage private developers to build housing "for sale and for rent" that targets state employees, workers and students who are "in need of housing" (Government of Vietnam, 2001, p. 1). In fact, the concept of "social housing" was first introduced in 2004 in the form of a "housing fund supported by state policies for rent or rent-to-buy to those within the 'policy categories' who face difficulty to improve their housing condition" (Decision 76/2004/QĐ-TTg, 2004, p. 2). Such a policy framework was subsequently substantiated with details regarding target groups, eligibility criteria and concrete measures to boost the housing supply.

The extension of target groups with stricter eligibility

The housing law in 2005 formally specifies social housing as "housing invested by the state, organization or individuals … for rent or rent-to-buy … with target groups including state employees, military officers and workers in industrial zones who are in need of housing" (Decision 76/2004/QD-TTg, Article 53). Coincidentally, the target groups were also priority groups for state housing in the central planning era. A complementary policy was introduced in 2009 for "housing for the low income" (Decision 67/2009; Resolution 18/NQ-CP), which extended the eligibility for 'Low-income housing' from state employees to include also low-income households in the private sector in urban areas as well as students and industrial workers. The most recent housing law, effective in 2015 (Decision 65/2014/QH13), further extends the coverage of social housing to include "state supported housing directed to those categories that are entitled to housing support according to this Law". Recent additions to the target groups include poor urban households living on social welfare, single elderly people, relocated households, and poor households in rural areas (Decree 188/2013/NĐ-CP).

While the target groups were expanded, the eligibility criteria became increasingly more stringent and exclusive. Initially, housing need was the only criterion for social housing. Applicants now have to be tenants who lived in over-crowded housing (as specified in housing law in 2005 and 8 square metres per person in Decree 188/2013/NĐ-CP). Income was later included as an additional criterion in 2009 with low-income housing being introduced as another social housing category (Decision 67/2009/QD-TTg). Only households earning below the average income in the city are eligible (Circular 36/2009/TT-BXD). One feature common to all the housing schemes is permanent urban registration. Verification of income from the work place and from the ward on housing need is required (Decree 188/2013/NĐ-CP, p. 19). In essence, this excludes workers in the informal sector and unregistered migrant workers who are not able to secure the necessary certification from the formal sector. This requirement excludes a very large segment of the urban population who are in need. Similarly, the requirement of a 20 per cent down payment for the rent-to-buy housing scheme (Decree 188/2013/NĐ-CP, p. 19) also imposes a serious hurdle for low-income households.

Mobilizing the corporate sector to produce social housing

Concomitant with a new strategy in the expansion of social housing is a shift in the finance of social housing development, from solely dependent on the state budget to involving private capital as a supplement. While the housing law of 2005 still emphasized the state's role in social housing provision, the adoption of a "socializing approach" also occurred (Decree 18/NQ-CP 2009, item III.1). Between 2009 and 2013, incentives like an exemption from land use rights fees, preferential VAT rate, corporate tax, assistance for site clearance, reimbursement for infrastructure outside of the projects and allowance for higher construction and land-use ratio have been introduced to increase private sector involvement (Decision 67/2009/QD-TTg; Decree 188/2013/NĐ-CP; Resolution 18/NQ-CP 2009). Further preferential policies were provided to social rental housing developers to allow the sale of rented apartments to sitting tenants after 5 years (at prices set for social housing at the selling time). To further boost the development of social housing, a "Twenty Percent Scheme" was introduced that

Figure 11.5 Social housing in Viet Hung new urban area, Hanoi.
Photo: Hoai Anh Tran.

requires all new development in new urban areas to set aside 20 per cent of their production for social housing (Decree 188/2013/NĐCP, Article 6–2a). To compensate the developers, 20 per cent of the total floor space of the "Twenty Percent Scheme" can be deployed for commercial purposes (ibid., Article 12–11). At the same time, a profit cap of 10 per cent was imposed on social housing projects to control social housing prices (ibid., Article 15–2a). Yet private developers seem not to be convinced. Very little social housing has in fact been produced through the "Twenty Percent Scheme" (Tran, 2015; World Bank, 2011).

In theory, individuals or households who intend to undertake social housing construction are able to join the incentive schemes. However, the eligibility criteria of such schemes are designed to suit medium to large companies. This essentially excludes small-scale development by individuals and households.

The Thirty Trillion package

The Thirty Trillion VND[2] package is another state program to offer incentives to encourage the production of housing for the middle- and low-income segments. It offers both mortgage loans to home buyers and loans for developers (World Bank, 2015). It was launched in the aftermath of the economic downturn in 2013 as a measure to revive the economy and to invigorate the stagnant real estate sector. To be eligible, developers have to invest in the production of social housing (with plans approved by the local authority) and equity of at least 30 per cent of the investment cost. Small-scale, incremental developments of social housing by individuals and households are also eligible. Eligible home buyers have to be state employees or low-income individuals who have no house or own a house of less than 8 square metres living space per capita and have paid social insurance for at least one year in an urban district. A down payment of 20 per cent is required, and the maximum repayment period is 15 years. The package has indeed incentivized several developers to change part of their commercial housing production to social housing. By 2015, the package had supported 45,000 households in mortgage loans and financed the construction of 65,000 affordable housing units (World Bank, 2015, p. 50). It was also one of the few policies that provided support to the demand side. However, the requirement of a 20 per cent down payment excluded low-income households.

The neglected renters

For many years after *Doi Moi*, the need of renters has been totally underestimated. With old public housing being sold to sitting tenants and the development of rental housing halted, only new housing produced in the market is for sale. The shortage in rental housing has intensified when escalating urbanization has accelerated migration to the city and hence pushed up the demand for rental housing.

However, the various strategies to encourage the production of low-income and social housing as well as the Thirty Trillion package are all directed at housing for sale. In theory, the gap in rental housing supply should have been met by housing for rent in the "Twenty Percent Housing Scheme", yet most of the housing produced in the scheme ended up on the sale market, apparently for cost-recovery considerations (*Vietnamnet*, 2016). After 2010, policy has been diverted to produce more rental housing. Yet disappointingly, only a small number of rental housing units have been built for industrial workers and students. As of 2014, of the 44 low-income housing projects, 10 worker housing projects and 13 student housing projects that have been planned in Hanoi, only 9 low-income housing projects, 4 worker housing project and 4 student housing projects have actually been completed (Pham, 2014). In fact, policy on rental housing has been inconsistent. While the housing law of 2005 set the priority of public housing development as "rent, rent to buy and for sale", actual implementation of this policy has instead placed the priority on social housing as "for sale, for rent and rent to buy" (Government of Vietnam, 2013).

Hence, prompted by such policy ambivalence, it is not surprising that developers tend to choose to develop social housing for sale and push rental housing production to a minimum. In Hanoi, only two social housing projects of rental apartments developed by corporate developers were completed in 2015 and 2016; they offer nominal 270 and 86 rental apartments, respectively (*Tienphong.vn*, 2016).

Rental housing has been given more attention in recent policies such as the Housing Strategies towards 2020, A Vision to 2050, and the new housing law 2015. The Housing Strategies emphasize the need to push forward development of rental housing, both through the state budget and state-supported private efforts (Prime Minister, 2011, p. 2). However, whether these efforts will lead to increased rental production is still unclear.

Affordability

After two decades of housing reform, it appears that Vietnam has produced an excess of up-market housing that is out of reach of ordinary households. The average price of such high-end housing stands at 25 times the average annual income of a household, which is substantively higher than the norm of 2 to 4 times often found in other countries (World Bank, 2015). At the same time, the supply of affordable housing has lagged behind demand. Even the so-called social housing units for sale are not easy to afford for the majority of low-income households. For instance, a "low-cost" apartment of 50 square metres was sold at VND 500–600 million, which is already a whole year's income for households at the lowest income quintiles. The World Bank study in 2014 found that the cheapest housing segment is affordable for households of the three middle-income quintiles (Q2, 3 &4) and medium-priced housing is affordable for those of the highest quintile. While very few urban households (8 per cent) can afford high-end housing (World Bank, 2015), households at the lowest income quintile can afford nothing in the market.

Apparently, policies that support ownership clearly do not benefit low-income households. Even households who earn an average income need to save all their income for at least 5 years in order to pay the 20 per cent standard down payment required for home buying. It is thus not surprising that households of the lowest income quintiles, who are already over-stretched, would have no capacity to save for housing purchase (World Bank, 2015, p. 16). The need for rental housing is equally critical, with high demand among the urban poor (UN Habitat, 2014). It is particularly acute for non-registered migrants who are barred from access to the formal housing sector.

AFFORDABLE HOUSING PROVISION BY THE INFORMAL SECTOR

Compared to the disappointing outcomes of the formal housing sector, which has full backing by the state with substantial financial support, the informal sector has been much more effective in providing affordable housing for social groups that are neglected by the formal sector: lower-income households, students and migrants. A wide variety of cheap apartments for rent and for sale are made available for such groups.

This dynamic private market can be clearly observed over the Internet with a great number of specialized websites for both renting and sale, for an apartment or even a room (e.g. *Phongnhatro* [rental room][3]) and *thuechungcunimni* (mini apartment for rent[4]).

A popular form of rental business is to subdivide an apartment into separate rooms for rent; owners can ask for a monthly rent of 1.2 million VND (*News.zing.vn*, 2015). Those with their own toilet facilities get somewhat higher rents. An even more popular form of low-cost housing solution is the *chung cu mini* (mini apartment blocks), tiny walk-up apartment blocks of 5–6 floors with 2 to 6 rooms on each floor built on small plots of land of 200–300 square metres in back lanes, where the land is cheap (Figures 11.6a & 11.6b). These are popular among households in the bottom two quintiles; most of them are migrants, students or young couples, with cheap rent at 700 to 1,000 million VND per month for a full apartment and convenient location in the inner city (Anh & Long, 2009). Owner-occupied alternatives are also available. Construction costs of such buildings are modest, and hence investors are often able to secure a profit margin of almost three times the investment (Anh & Long, 2009).

The mini-apartment for sale was a typical example of fence-breaking activities that were subsequently legalized. In fact, apartments smaller than 45 square metres for sale were illegal before 2010 as the then-housing law stipulated a minimum floor space of 45 square metres (a smaller size was allowed only for social housing). With *chung cu mini* that apparently defied regulations (and hence had no legal certification of ownership) already widely built, the government finally backed down in 2010 and rectified their existence. Degree 71/2010 allows the construction of "independent apartments in an individual house", and individual owners can build apartment blocks with two independent apartments of at least 30-square-metre floor space with separate entry and their own

Figure 11.6a Some Chung Cu Mini in Hanoi, 2017.
Photo: Vu Minh Anh.

Figure 11.6b Some Chung Cu Mini in Hanoi, 2017.
Photo: Vu Minh Anh.

toilet facilities. This was *exactly* how the "illegal" *chung cu mini* were built. Mini apartment buildings that already existed before the legislation were granted, retrospectively, full ownership status that allowed the resale of such apartments. The legalization of the *chung cu mini* caused an intense debate in the media. Corporate actors criticized the policy as being biased to individual households and small entrepreneurs and jeopardizing their interests. Supporters of the policy welcomed the move as it diversified players in the market and "now every household can be a developer" (*Plo.vn*, 2010); as a result, more low-cost housing could be produced to meet the demand for low-income households. A backlash for the *chung cu mini* owners came in 2015 with Decree 99/2015, which specified that ownership rights of flats in a *chung cu mini* will be granted only if the building was built legally (afforded a building permit) and complied with prevailing technical and safety requirements (*Vnexpress.net*, 2016). This decree implies serious administrative hurdles for those who want to invest in *chung cu mini* for sale.

While the informal sector makes significant contributions in providing affordable housing, especially rental housing for the poor, the poor living conditions found in this sector are worrying. Rental housing units created by subdividing an existing apartment lack both the appropriate amenities and privacy. In some cases, the rental unit is only a bed in a room. In many cases, tenants are exposed to unreasonable restrictions or unfounded extra charges from the landlords (*News.zing.vn*, 2012). Despite the policies that protect the rights of tenants, the lack of effective enforcement makes such policies almost obsolete.

Discussion: the duality of housing policy

Neoliberal-developmental hybrid approach

Vietnam's housing policies after *Doi Moi* clearly exhibit neoliberal features. Market imperatives are emphasized with an increased reliance on private capital for housing delivery and urban infrastructure. Housing is nearly completely commodified, and master-planned housing development by the corporate sector is promoted. There are also many supply-side supports to housing developers. To attract investment from private capital, large pieces of land are being leased cheaply to investors in exchange for their commitment to produce housing and infrastructure, and housing policy is apparently skewing towards 'profit making values at the expense of use values, social needs and public goods' (Peck, Theodore & Brenner, 2013, p. 1092). Yet this skew only creates opportunities for the investors to make windfall profits from commercial housing targeting the middle class while ignoring the need for social housing.

Despite the neoliberal outlook, this is no simple case of the victory of the market and the retreat of the state. The directives of master-planned housing development seek instead to reaffirm the role of the state in urban planning and housing production and increase state control of the housing market, something the state did not manage to do with the expansion of the incrementally produced popular housing during the 1990s (Tran, 2015). It means a deep involvement of the state via land allocation, via governmental approval of the development plan, via the various financial, technical and infrastructure supports.

Moreover, the state not only steers and controls urban housing development through investment and planning regulations, but is also directly involved – via entrusted equitized companies – in the implementation. The majority of large infrastructure and housing projects were carried out by equitized state companies and corporations in which the state holds more than a 50 per cent share. Those private companies that are entrusted with large-scale projects are those who have close connections with the government (Gainsborough, 2010; Tran & Yip, forthcoming). The legacy of the socialist system prevails, with state policies still skewed towards state control in relation to local government, and state-owned companies in relation to private ones, despite the rhetoric of decentralization and privatization and despite demands for a level playing field from provincial and local authorities and private firms (Malesky, 2004).

However, this situation cannot simply be interpreted as the return of state control as in the socialist era. Urban development is considered a strategic field of "new business", the leading players of which are capitalist companies that are "linked to state enterprises and/or bureaucratic institutions of the party state ... [with the aim] to exploit commercial opportunities that emerged during the reform years" (Gainsborough, 2010, p. 34). It is the apparent intention of the state to control the market as well as to be able to reap the profit (see also Shin, 2009 for a similar discussion on China).

The development of the post-*Doi Moi* housing policies presented above is a hybrid and seemingly self-contradicting process. On the one hand, fully-fledged neoliberal measures have put market forces as the imperative in boosting housing production, and commodification of housing is intensively promoted. On the other hand, the state is also trying hard to exert control over housing to fulfil its socialist missions, and hence the development of social housing has been high in the political agenda, albeit not high on the action plans. Hence, seemingly contradicting forces coexist: the neoliberal goal in facilitating the housing market for investors is promulgated alongside measures to increase state control over urban housing development. The result is a hybrid approach trying to achieve both goals. Hence the promotion of commodification of housing is integrated with the further development of social housing. This is in fact a further illustration of the hybridity of Vietnam's path-dependent reform trajectories, which have been extensively discussed in the literature (Gainsborough, 2010; McGee, 2009; Painter, 2005). Nonetheless, the formulation of social housing policies and favours to state employees manifests the deep-rooted socialist doctrines that echo the ideology of the command economy. While the policy of "low-income housing" was a significant improvement towards the adoption of a multi-sector economy, the additional requirement of urban registration still reflects the legacy of a socialist administrative and political mentality of state control over the population.

State-led and people-led

Vietnam's transition to a market economy can be described as simultaneously top-down and bottom-up. Whilst the party state still attempts to dominate the steering role via political mobilization and a monopoly of the decision-making process, "real policies" are instead increasingly shaped by "fence breaking" activities by local actors who challenge the official line and exploit the rules beyond their limits (Gainsborough, 2010; Masina, 2012; Painter, 2005). Hence, it is not uncommon to find national policies that are formulated at the central level being subtly "twisted" by various national and local authorities in implementations that may deviate from their original intention. Behind such processes is iterated bargaining, which involves interactions and negotiations between different fractions of the state as well as between the state and non-state institutions, communities, social groups and individuals (Painter, 2005). Very often, these "actually existing" policies would in turn loop back to the national level and be incorporated and rationalized into the revised national frame (Masina, 2012).

Consequently, such complex processes lead to uncertainty in the regulatory framework and gaps in the monitoring infrastructure. This problem encourages rent-seeking behaviour and breeds informality (Painter, 2005; Tenev, Carlier, Chaudry & Nguyen, 2003, pp. 14–16). Yet, informality can also emerge "from within the state" (Painter, 2005, p. 268). In fact, "illegal" renovation activities and housing transactions that were carried out in the state housing areas in the central planned era were in fact the first batch of "illegal" acts that were legalized after the economic reform.

Along a similar line, McGee (2009) argues that Vietnam's urbanisation is simultaneously a state-led project and a people-led project. Urbanisation involves complex processes of urban space production being shaped by negotiation, resistance and compromises between the main driving impulses of the state, the entrepreneurial sector and the popular sector (McGee, 2009, p. 235). The development of housing, especially affordable housing in Vietnam, illustrates this double process. On the one hand, state-led housing development in the forms of large scale high-rise development in master-planned new urban areas fits the state's vision of a modern Vietnam with the upholding of state control and modernization impulses (McGee, 2009) of Vietnam's housing policies. The state's support to boost corporate sector involvement could also be seen as an investment in new business opportunities that are of strategic importance for the state, both politically and economically (Gainsborough, 2010). Not only is the success of the formal sector a matter of national importance as it forms the main source of state revenue (Painter, 2005, p. 272), but it also helps in enhancing the image of the state (Gainsborough, 2010).

On the other hand, the majority of new urban housing was instead produced by individuals, households and small entrepreneurs outside the formal sector. This majority is particularly paramount in the provision of affordable housing alternatives that meet the demand of a great number of urban poor. However, this sector has been, for many years, undermined by the state. A dual strategy has been employed which, on the one hand, employs stringent regulations to suppress the sector and, on the other hand, turns a blind eye to informal housing production activities with the intention to lean on this massive force to supplement the formal sector in producing affordable housing.

A side effect of the failure of the formal sector in producing adequate affordable housing as well as a laissez-faire attitude towards the informal sector would have reinforced the image of a weak state capacity, which had already been conveyed in the visible images of out-of-control illegal extension and scenes of chaotic street vending. However, as argued by Gainsborough (2009, 2010), widespread informality, as well as legal ambiguity, are not necessarily signs of weak state power but rather expressions of an instrument of indirect government.

In this respect, the view that the formal and informal sectors are mutually exclusive may be misleading. The changing legal status of the *chung cu mini* clearly illustrates the shifting nature of what is considered "informal" and "formal", "illegal" and "legal". On the one hand, while the permissive attitude about informal activities and legalisation of some activities could be perceived as evidence of the government's submissiveness to grassroots demands and actions, it could also be interpreted as a form of "calculated informality" – selective enforcement of laws and partial authorization of what has been classified as unauthorized (Roy, 2005). In fact, legal ambiguity creates space for the local governments and officials to interpret the laws and regulations to their advantage (Tenev et al., 2003). Hence, the informal sector is not an "unregulated" sector but a form of "deregulation" in which the state is highly active and has the power to determine what is legal and what is not (Roy, 2005). The state can thus be seen as an informalizing entity from above, and informality is "an integral part of the territorial practices of state power" (Roy, 2011, p. 84).

Conclusion

Housing policies in transitional Vietnam manifest a dual nature: clear neoliberal features in pursuit of the commodification of housing co-exist with an enhanced state control of housing production with explicit promotion of social housing as well as a state-supported, corporate-led formal sector that produces only a small share of housing alongside a large but unacknowledged informal sector that provide the majority of urban housing. Vietnam's formal housing sector faces serious problems of housing affordability with an acute shortage of housing for the low-income and the poor. The housing problem could have been much worse if the dynamic informal sector had been unable to develop creative and low-budget housing alternatives.

The state's ambivalence towards the informal sector reflects the clash between ideology and instrumentality. One the one hand, the ideological baggage of state socialism drives the state in developing social housing for the socialist supporters; on the other hand, the neoliberal mentality in upholding market supremacy seems to compel it to promoting market provision of housing. At the same time, the need to monopolise power by the party state reinforces the continuation of the production of a coherent state in which the government is in full control, but the instrumentality leads to the tolerance of informal housing growth as viable solutions to the housing problems. It is the selective adoption of socialist doctrines and discretionary implementation of features of market mechanisms that help to maintain stability in times of change (McGee, 2009; Painter, 2005). However, as we can see in this chapter, some features of the socialist legacies, for example, the idea of the state's dominance of the economy, prove to be counter-productive to the development of a diverse and effective housing sector.

Notes

1 The nature of the "informal sector" is such that figures are difficult to obtain. The figures for the regional distribution of the informal sector are unavailable. National survey figures in 2007 show that the informal economy accounts for 23.5 per cent of total employment in Vietnam, which mean nearly half of all employment if agriculture is excluded (Cling, Razafindrakoto & Roubaud, 2011, p. 17).
2 1 VND = 0.000044 USD (as of November 2017).
3 See *Cho Thuê Nhà Nguyên Căn, Cho Thuê Nhà Riêng Tại Hà Nội* website (Full apartment for rent, private rental in Hanoi), 2018.
4 See *Thue chung cu mi ni* website (Chung cu mini for rent), 2018.

References

Anh, H. & Long, X. (2009, 21 July). Sot chung cu mini (The mini apartment fever). *Vneconomy.vn*. Available at http://vneconomy.vn/bat-dong-san/sot-chung-cu-mini-20090721121355429.htm, accessed on 22 October 2017.
Evertsz, H. (2000). *Popular housing in Hanoi*. Hanoi: Cultural Publishing House.
Cling, J. P., Razafindrakoto, M. & Roubaud, F. (2011). *The informal economy in Vietnam*. Hanoi: International Labour Organisation.
Gainsborough, M. (2009). Privatisation as state advance: Private indirect government in Vietnam. *New Political Economy*, *14*(2), 257–274.
Gainsborough, M. (2010). *Vietnam rethinking the state*. Chiangmai, Thailand: Silkworm Books.
Geertman, S. (2007). *The self-organizing city in Vietnam; Processes of change and transformation in housing in Hanoi*. PhD Thesis, Eindhoven University of Technology, The Netherlands.
Gough, K. & Tran, H. A. (2009). Changing housing policy in Vietnam: Emerging inequalities in a residential area of Hanoi. *Cities*, *26*(5), 175–186.
Government of Vietnam. (2001). Decree 71/2001/NĐ-CP on the priorities in investment in housing for sale and for rent.
Government of Vietnam. (2006). NĐ 02/CP Decree on the regulations of new urban areas.
Government of Vietnam. (2009). Resolution nr. 18/2009/NQ-CP on the regulation frameworks and policies to push forward the development of housing for students and housing for workers in industrial zones and low income in the urban areas.

Government of Vietnam. (2010). Nghi dinh: Qui dinh chi tiet va huong dan thi hanh luat nha o (Decree: detailed regulation and guidelines for the implementation of the housing law) (71/2010/NĐ-CP).

Government of Vietnam. (2013). Decree 188/2013/NĐ-CP of the government on the development and management of social housing.

Government of Vietnam (2015). Decree 99/2015/NĐ-CP Detailed specifications and implementation guides to the Housing Law.

Hanoi People's Committee. (2001). Decision 123/2001/QĐ-UB on the regulations on the investment and construction of the new urban areas, on housing renovation and repair in Hanoi City.

Harms, E. (2012). Neo-geomancy and real estate fever in postreform Vietnam. *Positions*, *20*(2), 406–436.

Harvey, D. (2005). *Neoliberalism: A brief history*. Oxford, England: Oxford University Press.

Higgs, P. (2003). Footpath traders in a Hanoi neighbourhood. In L. B. W. Drummond & W. Thomas (Eds.), *Consuming urban culture in contemporary Vietnam*. London and New York: RoutledgeCurzon, pp. 73–88.

Koh, D. W. H. (2006). *Wards of Hanoi*. Singapore: Institute of South East Asian Studies.

Housing Law. (2005). Law nr. 56/2005/QH11, issued on November 29, 2005.

London, J. (2009). Viet Nam and the making of market-Leninism. *The Pacific Review*, *22*(3), 375–399.

Malesky, E. (2004). Leveled mountains and broken fences: measuring and analysing de facto decentralisation in Vietnam. *European Journal of South East Asian Studies*, *3*(2), 307–336.

Masina, P. (2012). Vietnam between developmental state and neoliberalism. In K. S. Chang, B. Fine & L. Weiess (Eds.) *Developmental politics in transition: The neoliberal era and beyond*. Hampshire: Palgrave Macmillan, pp. 188–210.

McGee, T. G. (2009). Interrogating the production of urban space in China and Vietnam under market socialism. *Asia Pacific Viewpoint*, 50(2), 228–246.

Ministry of Construction. (2009). 16/11/2009 Circular 36/2009/TT-BXD implementation guidelines for the sale, lease and lease to buy and the management of low income housing in the urban areas.

News.zing.vn. (2012, 20 September). Nha tro va nhung "dieu kien" la doi (Rental housing and the strange "conditions"). *News.zing.vn*. Available at https://news.zing.vn/nha-tro-va-nhung-dieu-kien-la-doi-post276216.html, accessed on 21 October 2017.

News.zing.vn. (2015, 23 November). Kinh doanh nhà trọ không caần là chủ nhà ở Sài Gòn (No need to be a home owner in Saigon). *News.zing.vn*. Available at https://news.zing.vn/kinh-doanh-nha-tro-khong-can-la-chu-nha-o-sai-gon-post601969.html, accessed on 25 October 2017.

Nhà Riêng Tại Hà Nội (2018). Nhà Riêng Tại Hà Nội (Full apartment for rent, private rental in Hanoi). Available at: http://phongnhatro.com/cho-thue-nha-nguyen-can/ha-noi.html., accessed on 7 February 2018.

Nonini, D. M. (2008). Is China becoming neoliberal? *Critique of Anthropology*, *28*(2), 145–176.

Painter, M. (2005). The politics of state sector reforms in Vietnam: Contested agendas and uncertain trajectories. *The Journal of Development Studies*, *41*(2), 261–283.

Peck, J., Theodore, N. & Brenner, N. (2013). Neoliberal urbanism redux? *International Journal of Urban and Regional Research*, *37*(3), 1091–1099.

Pham, T. S. (2014). *Nhà ở xã hội tại Việt Nam: quan niệm, chính sách và thực tiễn (Social housing in Vietnam: concepts, policies and reality)*. Available at http://ashui.com/mag/chuyenmuc/bat-dong-san/10853-nha-o-xa-hoi-tai-viet-nam-quan-niem-chinh-sach-va-thuc-tien.html, accessed on 22 October 2017.

Plo.vn (2010, September 11). *Cong nhan "chung cu mini": Hop thuc hoa su da roi (Acceptance of the Mini apartment: Legalized What Has Been Done)*. Available at http://plo.vn/thoi-su/cong-nhan-chung-cu-mini-hop-thuc-hoa-su-da-roi-174385.html, accessed on 22 October 2017.

Prime Minister. (2004). Decision 76/2004/QĐ-TTg on housing development directive to 2020.

Prime Minister. (2009). Decision 67/2009/QĐ-TTg 24/4/2009 on the regulation framework and policies for housing development for low income people in the urban areas.

Prime Minister. (2011). Decision 2127/2011/QĐ-TTg national housing development policy for the year 2020 with the vision to 2030.

Quinn, B. J. M. (2002). Legal reform and its context in Vietnam. *Columbia Journal of Asian Law, 15*(2), 221–290. Available at https://papers.ssrn.com/sol3/papers.cfm?abstract_id=995599, accessed on 18 November 2007.

Roy, A. (2005). Urban informality: Toward an epistemology of planning. *Journal of the American Planning Association, 71*(2), 147–158.

Roy, A. (2011). Why India cannot plan its cities: Informality, insurgence and the idiom of urbanization. *Planning Theory*, 8, 76–88.

Schwenkel, C. & Leshkowich, A. M. (2012). How is neoliberalism good to think Vietnam? How is Vietnam good to think neoliberalism? *Positions, 20*(2), 380–401.

Shin, H. B. (2009). Residential redevelopment and entrepreneurial local state: The implications of Beijing's shifting emphasis on urban redevelopment policies. *Urban Studies, 46*(13), 2815–2839.

Tenev, S., Carlier, A., Chaudry, O. & Nguyen, Q. T. (2003). *Informality and the playing field in Vietnam's business sector*. Washington, D.C.: International Finance Corporation, World Bank.

Thue chung cu mi ni (2018). Thue chung cu mi ni (Chung cu mini for rent). Available at http://thuechungcumini.vn., accessed on 7 February 2018.

Tienphong.vn (2016, 5 July). "Chay" nha o cho thue (High demand for social housing for rent). *Tienphong.vn*. Available at www.tienphong.vn/Kinh-Te/chay-nha-o-xa-hoi-cho-thue-870700.tpo, accessed on 22 October 2017.

Thanh Uy Ha Noi (Communist Party of Hanoi). (1998). *CTr12/UBND Chuong trinh phat trien nha o Hanoi den nam 2000 va 2010* (Hanoi housing development plan for the years 2000 to 2010).

Tran, H. A. (2015). Urban space production in transition: The cases of the new urban areas of Hanoi. *Urban Policy and Research, 33*(1), 79–97.

Tran, H. A. & Dalholm, E. (2005). Forward owners, neglected tenants: Privatisation of state owned housing in Hanoi. *Housing Studies, 20*(6), 897–929.

Tran, H. A & Yip, N. M. (forthcoming). Neoliberal urbanism meets socialist modernism: Vietnam's post reform housing policies and the new urban areas of Hanoi. In Y. L. Chen, S. Asato & H. B. Shin (Eds.) *Contesting urban space in East Asia: Recasting neoliberalism upon housing*. New York, USA: Palgrave Macmillan.

UN Habitat. (2014). *Vietnam housing sector profile*. Hanoi: UN Habitat.

World Bank. (2011). *Vietnam urbanization review: Technical assistance report*. Washington, D.C.: World Bank.

World Bank. (2015). *Vietnam affordable housing. A way forward.* Washington, D.C.: World Bank.

Yip, N. M. & Tran, H. A. (2008). Urban housing reform and state capacity in Vietnam. *The Pacific Review, 21*(2), 189–210.

Vietnamnet. (2016, 12 July). Nha o xa hoi: mua da kho, thue con kho hon (Social Housing, Difficult to Access to Buy, Even More Difficult to Rent). *Vietnamnet.* Available at http://vietnamnet.vn/vn/bat-dong-san/thi-truong/nha-o-xa-hoi-315217.html, accessed on 22 October 2017.

Vnexpress.net. (2016, 10 June). Co duoc cap so hong khi mua chung cu mini (Can an "ownership right" be granted for buyers of flats in Chung Cu Mini). *Vnexpress.net.* Available at https://vnexpress.net/tin-tuc/phap-luat/tu-van/nen-lieu-mua-chung-cu-mini-chua-co-so-hong-3417115-p2.html, accessed on 22 October 2017.

12 Housing segmentation and diverging outcomes in housing wellbeing in Bangkok, Thailand

Thammarat Marohabutr

Introduction

Thai housing development has led to a housing shortage because of the increasing housing needs and excess housing demand from large scale rural-to-urban migration to Bangkok, the capital city, where the country's economic opportunities concentrate. This labour force has supported Thailand's economic growth based on export-oriented industrialisation depending on labour and semi-skilled intensive mass production. This growth has contributed to the current nominal gross domestic product per capita of around 5,500 US dollars and 17,000 US dollars in terms of purchasing power parity, pushing the country's status to an upper middle-income economy. The National Housing Authority (NHA) was established in 1972 to enhance housing opportunities for low-income groups and solve aggravated slum problems caused by low-income people and migrant workers who had no choice other than setting up informal settlements. Formal public housing projects were built alongside other minimal options, including a slum-upgrading scheme. However, the NHA has not become the mainstream housing provider for people as its role has been weakened whereas the role of private-sector housing has become robust.

In the mid-1980s, private housing activities in Bangkok flourished, contributing to dynamic housing investment, especially in the middle- and high-income segments. Nonetheless, the private housing market collapsed because of the economic crisis in 1997. By the late 1990s, robust housing investment by the private sector had, however, already promoted housing opportunities for Bangkok's population and dominated the housing market. After the 1997 crisis, the government-initiated policies and measures to revive the collapsed housing market. With the revival of the housing market, Bangkok's people regained affluence and the ability to afford private-sector-built formal housing. In term of housing for the low-income segment, pro-poor housing provision by the NHA was revitalised at the beginning of the 2000's. It comprised the Baan Eua Arthorn low-cost formal public housing project and the Baan Mankong slum-upgrading scheme. In the aftermath of the crisis, the revival of housing activities therefore generated different levels of housing wellbeing for Bangkok's urbanites across the two major segments of housing, the middle- and high-income segments and low-income segment. This chapter elaborates on the development of contemporary housing policy in Thailand and discusses the housing wellbeing of Bangkok's people residing in both segments, with the attending social implications.

Development of contemporary Thai housing policy

The early era of concrete housing policy in Thailand can be traced back to the establishment of the NHA in December 1972. In response to the rapidly growing number of slums, its basic objective was to provide public housing for rent, lease-purchase or purchase; to subsidise citizens wishing to own housing or individuals wishing to provide public housing for rent, lease-purchase or purchase; to manage building or housing estate businesses; and to improve slum settlements (NHA, 1986). The NHA has operated as a business-like agency to implement its objectives. This style of operation means that the private sector could take part in some of the NHA projects, especially the projects for middle- and high-income groups (Chiu, 1984). This practice was translated into a reduced public role of the NHA in developing housing for the urban poor in the mid-1980s (Buracom, 1987).

The projects under the NHA's First Plan had to depend on government financial support heavily. However, the Plan had to be halted in 1977. For one thing, the housing projects were too expensive as the standards of construction had been set too high. Indeed, immense costs and rising prices of land and construction materials were the most enfeebling of all constraining factors. One possible solution was to lower standards so that more housing units could be constructed (Huwanan, Radomsuk & Iamnoi, 1991; Tanphiphat, 1983). The NHA was obliged to transform its policy towards more affordable and self-sustaining alternatives, including slum upgrading, emphasising on-site upgrading such as walkway improvement and provision of facilities like drainage, electricity, etc. Opposed by appointed businessmen and other NHA top administrators, the NHA therefore adopted a marketised housing approach by constructing and selling higher-priced housing units to the middle- and high-income groups, aiming at making profits. The NHA retained the profits and then used them for cross-subsiding the construction of low-cost units to be sold to low-income people at low prices (Buracom, 1987).

In the mid-1980s to 1990s, housing development became robust due to the assertiveness of finance capital alongside a high and double-digit GDP growth rate, favouring the role of private developers in the housing industry. During this period, housing finance was improved and became the source for developers to get funds through the operation of Government Housing Bangkok (GHB) and commercial banks. Not only developers but also house buyers could obtain loans for house purchases. In fact, there were several sources of credit. Both public and private banking facilities provided credit to developers and house buyers (Yap & Kirinpanu, 2000). This access enhanced the role of private-sector housing. Until the first half of the 1980s, the private sector could construct around 10,000 to 20,000 units a year (Foo, 1992a, p. 100). Because of their financial viability, private developers managed to build almost 58,000 housing units in 1989 (Foo, 1992b, p. 1139).

With housing actively developed by the private sector, the share of private housing surged massively and has now become dominant in the Thai housing market, accounting for over 74 per cent of the total housing stock since 2001. Since 1990–2010, the home ownership in Bangkok and five adjacent provinces, also called the Bangkok Metropolitan Region (BMR), however, has declined. In Bangkok, the home ownership rate dropped from around 61 to 50 per cent; it also fell from around 78 to 58 per cent in five adjacent provinces. This decrease happened alongside the surge in migrants to Bangkok and its suburbs, which made the total population of Bangkok grow from 5.9 million people to 8.3 million people, and the rise of living costs in these areas during the two decades (NSO, 1990, 2010). As the BMR has a greater number of urbanites than other parts of the country, the types of housing are more varied, with a greater proportion of townhouses and condominiums and a smaller number of detached houses (JICA, 2013). Instead of being an effective response to the housing demand of the low-income group, the rapid increase in the housing activities led by the private sector has been distorted to become a speculative means for the middle- and high-income people employed in the formal sector to gain more wealth, which emerged along with the economic boom (Pornchokchai, 2002, p. 6). According to an estimation by the GHB, a housing bubble loomed in the housing sector when about 300,000 housing units were left vacant for speculative purposes in the BMR in 1995, or around 40 per cent of the total constructed units. These vacant units accounted for an accumulation of the new housing stock to be produced in the market for two years (Leightner, 1999, p. 368; Phongpaichit & Baker, 2002, p. 174). As speculators began releasing units because of the fear of a surplus, purchasers also stopped paying their housing loans. Therefore, banks ceased to give loans as a large amount of accumulated unpaid debt remained unsettled by developers and borrowers (Yap & Kirinpanu, 2000). At the same time, currency speculators started to attack the Thai Baht,[1] coercing the government to float it, which meant its value quickly dropped. Therefore, loan borrowers in foreign currency in the housing sector faced consequential deficits in foreign exchange and could not pay back their loans (Yap, 2014). A large number of purchased housing units was foreclosed, but nobody could buy them. Consequently, the Thai housing market collapsed along with the 1997 economic crisis.

After the crisis, several measures were initiated to heal the collapse and stagnation of the housing and real estate sectors, including domestic measures to stimulate the demand and supply side. On the supply side, a more compromising approach between housing developers and lenders was regulated through the amendment of the Bankruptcy Act. Furthermore, property funds were introduced so that they could be used as capital sources and perform as equity financing for the housing sector. To promote housing demand, the government issued policies such as taxation privileges, transfer fee reductions, incentives for house buyers through the Government Pension Fund (GPF) and pension funds for state enterprises employees, including provision of special low-priced housing for people with unstable incomes. The goal of these measures was to encourage people to purchase houses (Vanitchvatana, 2007). In the public sector, the government announced an initiative called "A Million Housing Units" in late 2002 to solve the housing problems of the urban poor. The new subsidised low-cost public housing scheme was therefore founded under the Baan Eua Arthorn (which literally means "we care" housing) project in 2003 through the NHA's implementation. Informal housing through slum upgrading was also enhanced. The Baan Mankong (which literally means "secure" housing) slum upgrading was said to benefit people living in the slums by improving the physical conditions of their existing dwellings and establishing security of tenure, which had been neglected during the earlier period through means such as long-term land leasing agreements with land owners to eradicate eviction threats (Boonyabancha, 2005). Nonetheless, the revival of housing development in both public and private sectors has reaffirmed a two-tier housing segment in Thailand. While the private-sector housing has served the middle- to high-income groups, public housing has tended to cater to the poor. Different outcomes of housing well-being in both segments have clearly been substantiated in Bangkok, where housing activities are most vibrant.

Middle- to high-income segment

The saturation of Bangkok's housing development by the private sector within the inner-city limit was seen in the mid-1980s. Nonetheless, as the economy recovered during the second half of the 1980s, private housing development moved to suburban areas with the construction of new roads. In addition, because of financial restructuring and the promotion of housing loans through the GHB and commercial banks, new private housing estates, mostly townhouses and detached houses, were constructed on the fringes of Bangkok. On the demand side, housing finance became a fundamental source for house purchases along with the growth of the middle- and high-income classes (Kirinpanu, 1993; Tanphiphat, 1993). As housing development for the high-end market boomed, marketing competitions focusing on better locations, infrastructures and facilities were rampant among developers. The newly built housing estates were diverse in size and styles, highlighting symbols of modernity and good taste for exclusive classes. Neo-classical, Californian-Spanish, Tudor style or mixed style houses were built in estates, which were equipped with fitness and recreational facilities and exclusive clubs that fulfilled aspirations to elite status (Askew, 2002). A growing number of condominiums instead of detached houses and townhouses represents the contemporary direction of the housing pattern in Bangkok as population densities have increased in the city centre.

The share of people living in condominiums grew from 6.2 per cent in 1980 to 9.7 per cent by 1990, while the proportion of people living in townhouses, a popular housing option of the middle-income group, increased from 7.4 per cent in 1980 to 9.3 per cent by 1990 (NSO, 1980, 1990). An increase in land prices in the suburbs also led to more subdivision of living space. Consequently, more condominium buildings were constructed on the fringes to serve middle-income people. In addition, more low-cost townhouses were built in the mid-1980s as alternatives to condominiums (Chulasai, 1983, pp. 176–177; Kirinpanu, 1993, pp. 9–11). According to the latest national housing census, the share of townhouses stood at 35 per cent of total housing units in Bangkok while those of detached houses and condominiums were 32 per cent and 31 per cent, respectively (NSO, 2010). Because of an enabling strategy in the 1990s, the private sector was encouraged to develop more affordable housing units, for which the government initiated financial programmes to supply housing finances to the lower-tier market (Pugh, 1994). Although the enabling strategy has been discontinued since the beginning of the 2000s, the private sector has developed new housing projects for the lower-tier purchasers in the middle- to high-income segments because of the increase of housing loans for developers from commercial banks, surging from 4.1 billion Baht in 2011 to 5.9 billion Baht in 2016 (Bank of Thailand, 2017).

Living in upper-market housing: a money matter

The blistering growth of urban sprawl has been led by the growing number of projects for the middle-income group. Surroundings, the short distance to the workplace and convenient transportation are the key factors to determine where to build these projects. Unlike the case of housing for high-income people, price is the most important factor for middle-income people when purchasing houses (Hara, Hiramatsu, Honda, Sekiyama & Matsuda, 2010). As housing activities in the private sector are market-oriented, many developers compete for the middle-income homebuyers by improving the quality of housing units and curbing costs to yield competitive prices to purchasers (Pornchokchai, 2002). The recent decrease of government fees for purchasing houses is an important factor boosting purchasing power, increasing housing opportunities for Bangkok's people, especially the middle-income group. For instance, the ownership transfer fee was reduced from 2 per cent to only 0.01 per cent of the appraisal price. The property mortgage fee also decreased from 1 per cent to just 0.01 per cent. In addition, an income tax reduction was granted to house buyers paying up to 300,000 Baht. Also, the income tax was reduced from the interest payment of a housing loan up to 100,000 Baht (Real Estate Information Center, 2010).

The affordability of housing units, particularly those built by private developers, for Bangkok's upper class has improved greatly because of continuous housing finance policies. The Agency for Real Estate Affairs collects the data on house prices. It found that, because of the bubble before the economic crisis, house prices skyrocketed in 1996–1997 before plunging in 1998–2003. From 2003 to 2004, house prices increased again. In specific housing projects, the house prices for the middle- to high-income segments rose in all products in between 1994 and 2010. For instance, the prices of medium-cost condominiums increased from 1.1 to 2.5 million Baht while those of high-cost detached houses, the most expensive housing product, rose to 5.1 to 7 million Baht (Pornchokchai, 2010, pp. 58–62).

In the aftermath of the crisis, home loans were re-originated in 1999 to provide opportunities for purchasers to buy houses when the overall economic performance stabilised. Considered less menacing, commercial banks and the GHB competed to provide home loans with lower interest rates because the revived housing activities had reflected the real demand rather than speculation, as in the pre-crisis era. Specifically, the GHB tried to promote affordable home ownership in 2002 by offering the lowest mortgage interest rates, 6.25 to 6.75 per cent, compared to the rates offered by the commercial banks of 7.10 to 7.75 per cent. It also played a leading role in a 30-year loan term extension and an increased loan-to-value ratio of up to 90 to 100 per cent (Kritayanavaj, 2002, pp. 19–22).

Since 1995 the preferences for housing types for Bangkok people have significantly altered, with less preference for townhouses and detached houses, whereas condominiums have become a favourite option (JICA, 2013). According to the Agency for Real Estate Affairs, the share of townhouses and detached houses among total housing stock stood in 1995 at 35 per cent and 12 per cent, respectively, while that of condominiums was only 36 per cent. In 2010 the proportion changed with 22 per cent for townhouses and 14 per cent for detached houses, but the share of condominiums increased to 55 per cent (JICA, 2013, p. 18). The change has been caused by the change in family structures and the increase in income (JICA, 2013). That is, "families have earned more but lived apart" (Poapongsakorn, 2013). From 1990–2010, the average size of Bangkok's household steadily decreased from 4.3 persons to 2.7 persons. During this period, the proportion of single-person households also increased from 7.9 per cent to 23 per cent, and that of female-headed households surged from 28.7 per cent to 36.6 per cent (NSO, 1990, 2010). Household incomes also increased by about 70 per cent in Bangkok and 55 per cent in the BMR on the whole because of better and stronger economic performance. Nevertheless, the inflation rate grew by around 30 per cent, and the prices of detached houses rose by around 20 per cent. This trend may imply that Bangkok's housing has become more affordable. However, the major concern is that housing units built by private developers are for households with incomes more than 15,000 Baht a month[2] (Fernquest, 2013). The minimum daily wage also rose to 300 Baht a day or roughly 9,000 Baht a month since January 2013, so affording private sector housing is even more difficult for workers earning daily wages.

A recent study, the only currently available on the subject, estimates that Bangkok's households with an average income of 30,000 Baht per month are able to buy housing valued at about 1.25 million Baht. Currently, around 60 per cent of households in Bangkok earn up to 30,000 Baht a month, and these households can afford only 21 per cent of housing units. However, if the income threshold rises to 50,000 Baht per month, 80 per cent of households fall within this income level, and 59 per cent of housing units become affordable. In 2010, 42 per cent of housing units were valued at over 2 million Baht. However, just 20 per cent of households could pay this price. This fact attests to the fact that, in general, housing units in Bangkok tend to be unaffordable for many families. Rather, the units are built for middle- and high-income groups (JICA, 2013, pp. 18–21). The findings from this study correspond to the suggestion of a commercial bank recommending that people wishing to buy a house valued at 1.25 million Baht should earn at least 30,500 Baht a month to repay a 10-year housing loan. However, if they wish to prolong the loan term up to 30 years, they need to have a monthly income of at least 18,800 Baht (Kasikorn Bank, 2017).

Accessibility and security as amenity for living

As Bangkok has grown, it has become more densely populated, with an increase from 3,758 to 5,258 inhabitants per square kilometre from 1990 to 2010 (NSO, 1990, 2010). With a surge of newly built housing projects spreading to the suburbs, traffic congestion remains a notorious problem for Bangkok. In the absence of any efficient rail-based transport policy, road-based transportation has long been dominant (Rujopakarn, 2003; Tanaboriboon, 1997). It was estimated in 2009 that there were 6.1 million cars registered in Bangkok. This number accounted for almost one fifth of the total 27.2 million cars registered in the country (BMA, 2010). Bus transit, which is inefficient and insufficient, had been the only major mode of transportation serving daily commuters. The Bangkok Mass Transit System (BTS) or "Skytrain", the first mass transit system, began operation in the city centre in 1999 with two train lines and 20 stations. More mass transit railway lines are now in operation by three operators, namely, the Bangkok Transit System (BTS), the Mass Rapid Transit (MRT) and the Suvarnabhumi Airport Rail Link (SARL). A number of new lines with dozens of stations are being built to extend to the suburban areas, with planning and bidding processes for more lines underway. Mass transit infrastructure and increased income are key factors changing the living patterns of Bangkok's middle- and high-income groups, especially those living in condominiums, generating a "Skytrain" generation wishing to live and work as near as possible to mass transit stations (CB Richard Ellis, 2012). As the new mass transit lines have been built, linked, and extended to the suburbs, the expansion of the mass transit network has also led to residents living in housing estates on the fringes of Bangkok.

In terms of security, creating good, safe living environments is a government policy. Such a policy enhances public awareness and preserves the public's interest in ensuring livability. However, this policy has not succeeded in the housing market (JICA, 2013). Nonetheless, acceptable standards of public security and social rules supposedly to be ensured by government agencies are not uniformly achieved in some areas of Bangkok (UNICEF, 2009). Security in housing and neighbourhoods has become a crucial factor for middle- and high-income households, which have greater autonomy to buy houses or choose places to live. For instance, they tend to opt not to live adjacent to slums, where the problems of narcotics and robbery are prevalent (Yap, 1996). Housing with closed access and security guards has become common in mid- and high-end private developer-built projects. Housing estates with detached houses and townhouses are usually equipped with security check points and high walls (Askew, 2002). For instance, security is usually guaranteed in Bangkok's condominiums with round-the-clock security guards, closed-circuit televisions on most access points to buildings, internal televisions monitoring security and key cards allowing access of only people living in the building.

A juristic person as a means of managing common property

Although the trend of gated communities has not noticeably existed in Bangkok (Kramer, 2011), enhancing cohesion and the sense of community among better-off residents may be achieved through the mechanism of the Juristic Person, notwithstanding the fact that its prime objective is to protect customers' rights after sales of housing units built by private developers. The government passed the Condominium Act (No. 4) in February 2008, and it was put into effect in July 2008. This law has contributed to substantial changes in condominium management through the compulsory Condominium Juristic Person, which protects the rights of the condominium unit owners. Apart from the ownership of individual units to which the residents have full private rights, the Condominium Juristic Person is responsible for common property within the building compound that residents share. The residents are required to pay annual maintenance fees for this purpose. The common property includes, for instance, land on which the building is constructed; land provided for common use; the structure and construction of the building to protect the condominium against damage; the building or part of the building and equipment provided for common use; and machinery, tools, facilities and services provided for common use, etc. A manager of the Condominium Juristic Person is therefore appointed by a resolution at a general meeting of unit owners by a 25 per cent vote. The manager has powers and duties to act in the interests of the condominium by following the provisions or resolutions of the unit owners' general meeting or a committee. In addition, a Condominium Juristic Person committee consisting of representatives from the unit owners, together with its chairman and vice chairman, is appointed at a unit owners' meeting to control the management of the Condominium Juristic Person. Consideration of any matter within the condominium compound, especially the use of common property, must be decided and solved by majority voting on resolution at committee meetings.

The mechanism of the Juristic Person has been extended to housing estates as well. Appointment of the Housing Estate Juristic Person under the 2000 Land Allocation Act was instigated to make its responsibilities similar to those of the condominium. However, the implementation has not been effective, according to the intention of the Law on Establishment of a Housing Estate Juristic Person. Because the establishment of a Housing Estate Juristic Person is not compulsory, as in the case of condominiums, only a small number of Housing Estate Juristic Persons have been established compared to the total number of housing estate projects. The main reason is the limit of powers and duties provided by the law. In addition, house purchasers are unwilling to pay maintenance fees to support the management of the Housing Estate Juristic Person. Also, residents lack confidence that the Juristic Person's management, powers, duties and supervision could solve the problems of common living (Chairungrojsakul, 2009).

Low-income segment

During the early 2000s, 5,500 low-income communities in which 8.25 million people were living in insecure and poor conditions were scattered in about 300 cities. Further, 3,700 of these communities were facing land insecurity. Thirty per cent of people living there were squatters, whereas another 70 per cent rented land without long-term contracts. Four hundred and forty-five communities faced a risk of eviction, and 70 and 80 per cent of tenants could not afford formal housing (Boonyabancha, 2005, p. 22). During the campaigns for the 2001 general election, politicians delivered vigorous proposals for pro-poor programmes to enhance housing opportunities for the urban poor. After a new government had been elected, its ministerial board agreed to initiate two pilot pro-poor housing projects in July 2003: the Baan Eua Arthorn low-cost formal housing and the Baan Mankong slum upgrading scheme.

The Baan Eua Arthorn low-cost public housing

The Baan Eua Arthorn scheme is a low-income housing scheme based on lease-purchase under the responsibility of the NHA. It emphasises building new housing stock in the form of walk-up flats and small detached houses for the urban poor with household incomes of less than 15,000 Baht per month. Mortgages with monthly payments less than 1,500 Baht were to be provided. By contracting mainly to the private sector, the NHA projected to complete 600,000 units within 5 years (Khongpaen, 2003, p. 3). To match the actual housing demand, however, the NHA reduced the construction target to about 300,000 units by dividing the construction into 5 phases. The NHA planned to achieve 4,131 units and 7,100 units in the first and the second phases, respectively, of 2003. The target was increased to 71,300 units and 118,661 units in the third and the fourth phases, respectively, of 2004 and 2005. In the fifth phase of 2006, 99,312 units were to be achieved (NHA, 2008, pp. 1–1). In spite of the decrease, the Baan Eua Arthorn housing scheme never achieved its target in any phase. Table 12.1 indicates that the housing construction of the Baan Eua Arthorn project is emphasised in the BMR rather than regional cities. A political issue arose: the South never benefited from the Baan Eua Arthorn scheme

Table 12.1 Performance of the Baan Eua Arthorn scheme by phrase and location, 2003–2006 (units)

Phrase	*Location*				
	BMR	*Centre*	*North*	*Northeast*	*South*
1st phrase	3,535	–	596	–	–
2nd phrase	2,943	300	430	800	–
3rd phrase	41,753	3,444	2,564	9,391	–
4th phrase	20,312	9,160	6,425	5,654	–
5th phrase	40,233	3,727	1,755	1,742	–
Total	108,776	16,631	11,590	17,587	–

Source: NHA (2008, pp. 1–3).

because it was not the political base of the government. The NHA reported that more than half of the construction was implemented in Bangkok and its suburban provinces each year. In the BMR, most Baan Eua Arthorn units were built and located in fringe or suburban areas, where land is cheaper but transportation is inconvenient. During the late 2000s and the beginning 2010s, short-term governments have not abandoned the low-income public housing scheme. The Baan Eua Arthorn policy was, however, replaced by the Baan Pracharat scheme for the purpose of selling up the unsold 10,470 units built under the former Baan Eua Arthorn scheme. In addition, the NHA has planned to build 34,038 public rental housing units for low-income people in 2015–16 and an additional 35,000 units in 2017 (Ministry of Social Development and Human Security, 2017).

To maintain collaboration with the private sector, a subsidy of 80,000 Baht per unit is allocated to private contractors for construction costs of facilities such as community centres and expenditures for the first five years of maintenance, including utilities such as electrical fees for lighting and management work after completion. The income criteria for purchasing a unit in the Baan Eua Arthorn housing estate were, however, revised many times. In 2005, only families with incomes less than 17,500 Baht could purchase units. Implying less affordability for many poor, the income thresholds of eligible purchasers were again revised to less than 22,000 Baht in 2006 and 2007 and less than 30,000 Baht in 2008 (NHA, 2008). On the bright side, the Community Organizations Development Institute revealed that the scheme continuously covered the housing needs of the urban poor in 78,607 units in 1,287 communities in 70 provinces out of the total target by March 2009. When the NHA committee realised that the Baan Eua Arthorn scheme had consumed immense public expenditures of roughly 273,209 million Baht, the committee decided to end the scheme (NHA, 2008, pp. 1–1). The construction of on-going units was suspended. The Policy and Planning Department of the NHA revealed that 83,707 units of the Baan Eua Arthorn housing stocks had been completed in 2007. In addition, there was a critique about not yielding benefits to poor purchasers because the NHA had set the criteria too high. Banks did not approve mortgages for many poor people as the proposals did not pass such criteria. Furthermore, a number of housing units in the Baan Eua Arthorn project remain unsold as it cannot compete with housing projects by private developers and does not attract interest from the upper-income groups.

A survey of satisfaction conducted with the Baan Eua Arthorn residents found that living in Baan Eua Arthorn housing estates did not interest half of the respondents (Tanaphooma & Bart, 2015). Quality and standards of the units were the major concerns. The small living units were suitable only for small families while larger families faced difficulties in extending the unit areas to accommodate all members. The large plot of the project also affected lifestyles as it changed community living to more individual living (Supawittayanan, 2003). Furthermore, the residents complained about poor housing standards and environmentally dangerous materials and facilities such as roof tiles and ceilings, width of stairs, brick layer, roof structure installation, flooring condition, restroom design, pest control, household drainage pipe system, housing surrounding and waste management. Estate management, life and property security, common regulations and the availability of shops and parking lots were also concerns affecting their social lives (Verapreyagura, 2006).

The Baan Mankong slum-upgrading scheme

Whereas the Baan Eua Arthorn scheme was terminated, the Baan Mankong slum-upgrading scheme was initiated to improve the living conditions of slum dwellers by upgrading the existing conditions based on self-help and community development. Anticipating the real economic situations of the urban poor, less of the public budget is used to hastily solve housing problems in slum areas. The citywide slum-upgrading scheme at the national level is famous for emphasising community participation. Improvements in physical housing conditions, together with the engagement of poor residents in making housing policies for their own communities, are the merits of the scheme. Therefore, making slum-upgrading schemes successful provides beneficial lessons on the significance of community participation and collaboration among different actors (Bhatkal & Lucci, 2015).

During the late 1990s, extensive local development projects in various forms emerged in Thai society. These projects, such as credit savings or community businesses, were formed by the community networks within specific areas. They also evolved and extended to external networks linking the same socio-economic niche with other areas. The idea of solving housing problems under the aegis of a civil development process therefore emerged. It was believed that the concept of local development and community networks was synonymous with the initiation and enhancement of problem-solving for informal housing. With initiatives from the community organisations, people living in specific communities have participated and contributed to these projects in the roles of planner, implementer or financial administrator. All in all, these phenomena in Thai civil society could shift the role of a community from a development recipient to a development initiator. This self-help and community participation shift is regarded as the basic principle for implementing the Baan Mankong scheme.

The CODI and community participation in the Baan Mankong scheme

The Community Organizations Development Institute (CODI) was established in 2000 as an independent public organisation under the Ministry of Social Development and Human Security. Because of its independent administration, it can work more independently with greater possibilities and flexibility with wider connections and new possibilities to enhance co-operation and co-ordination among independent community organisations and networks nationwide. These strengths fit the objective of slum-upgrading based on self-help and community participation. The duty to implement the Baan Mankong scheme was therefore assigned according to the Ministerial Agreement in January 2003.

The Baan Mankong slum-upgrading project was founded to uphold the collaboration between low-income households and their community organisations and networks based on self-help and community participation to solve the housing problems of the urban poor. These community organisations and networks work with local government, professionals, academics and non-governmental organisations in their areas to survey all slum communities and then plan slum-upgrading projects to improve housing conditions. After the plans have been culminated, the CODI provides subsidies for infrastructure and housing loans to the communities. With self-management, the community's own determination and its dwellers' participation, the upgrading is to be implemented in existing communities. If relocation is required, new communities can be built on distant land, if necessary, or on nearby land to minimise the economic and social costs of households. Nonetheless, the Baan Mankong scheme will not stipulate unnecessary conditions as it aims to offer freedom for slum communities and their networks to designate their own projects. The challenge of the project is to promote slum-upgrading in ways that allow slum communities to lead the process and create local partnerships. In addition, as eviction threats have long been a major problem of slum communities, the scheme also promotes tenure security through negotiations settled locally on a case-by-case basis through any possible means such as co-operative land purchase and long-term lease contracts under construction approaches such as re-blocking and relocation (CODI, 2012).

Re-blocking is an approach emphasising community improvement on the existing site to improve the layout plans and existing infrastructure. In this approach, the existing community is not affected much. However, purchase of land *in situ* on a collective basis is considered if necessary. Greater security is the main benefit to the tenants as their community can be further developed as the land already belongs to them. For example, the residents of a community in Bangkok used this approach to upgrade their housing (Bhatkal & Lucci, 2015).

Although they had been threatened with eviction, they negotiated with the private land owner to buy the land where their community was located at market value. After setting up a co-operative, the tenants got a CODI loan to pay for the land purchase and upgrading. To reduce the cost per household, many tenants used materials from their previous houses to build new ones and gradually upgraded them. They also made agreements with municipal departments to bring electricity and running water to each house and delegated the construction to paid community members.

Relocation yields advantages to dwellers because more security can be achieved through housing development undertaken in a newly built community at a new location. However, the tenants have to adjust their lives as they need to relocate from the former community to the new site, which may be far from their jobs or schools. As a consequence, they may need to pay more for transportation to work or study. The trade-off is that they may develop their new community as they wish more freely, as tenure security is fully granted on the new site. A community close to the Port of Bangkok used the relocation approach despite the fact that it relocated to a new site adjacent to the original one (Boonyabancha, 2009). The community has occupied the land of the Port Authority of Thailand in Bangkok and faced eviction threats several times. The dwellers negotiated with the Port Authority to allow them to build a community with a 30-year lease on new land owned by the Port Authority and located one kilometre away. The community therefore took a loan from the CODI to relocate and built a new community.

In terms of financial support, the CODI provided an infrastructure subsidy of 25,000 Baht per family if upgrading was implemented on the existing land plots. If re-blocking was considered, a subsidy of 45,000 Baht would be granted. However, if relocation was needed, the CODI would provide a subsidy valued 65,000 Baht per family. The families could then apply for low-interest mortgages to pay back the subsidies to the CODI (Boonyabancha, 2005, p. 26). In addition, a grant equivalent to 5 per cent of the total subsidy would be given as a contribution to management costs of the community organisations or networks. The management of financing had to be communal, not according to the needs of any individual families or tenants.

The target of the Baan Mankong's initial plan was to achieve housing improvement and tenure security for 300,000 low-income households in 2,000 poor communities located in 200 cities within five years. This target represents at least half of the urban poor settlements in Thailand. In 2004, the projects of about 1,500 units were first implemented in ten pilot communities, six in Bangkok. Later, the target was extended to 20 other communities when the pilot projects succeeded. In 2005, upgrading of 174 slums in 42 cities of about 15,000 units was targeted. From 2006–08, the target was to build 285,000 units in 200 cities. The goal was actually achieved with 1,010 communities operating under 512 projects in 2008 (Boonyabancha, 2005, p. 25; 2009, p. 309). In 2009, it benefited 80,201 households in 1,319 communities. The communities located in the BMR have benefited the most, sharing the highest proportion of 376 communities and accounting for 33,428 households with an approved budget of 2,679 million Baht (Table 12.2). In 2012, the scheme benefited 91,805 households in 1,637 communities operating under 874 projects nationwide (CODI, 2012, p. 13). The scheme has been continuously supported and implemented by subsequent governments to the present. For instance, the Democratic government pledged itself to continue supporting the CODI on the Baan Mankong project by giving financial support valued at 1.5 billion Baht to promote security of tenure for low-income people living in the slum communities (*Krungthep Thurakij Online*, 2009).

Given support and steering action, the community organisations and their networks are the key actors in the Baan Mankong slum-upgrading scheme as they have granted the rights to control funding and management to the community and its dwellers. The organisations have been involved in most of the building, from planning to construction. This approach will fit the residents' demands, whether housing style or finance, more appropriately as they can decide everything by themselves. As the scheme grants flexibility in financing, residents can plan, implement and manage it directly. The government agency of the CODI is not the planner or implementer on a conventional top-down basis any longer (CODI, 2012). Therefore, the Baan Mankong scheme is more than a physical upgrading as every group of slum dwellers can take part in the housing development process without domination by others or exclusion from the development process (Posriprasert & Usavagovitwong, 2006). The community's decisions and management for the sake of their betterment will stimulate and enhance confidence among the urban poor in self-management and determination. With more embedded managerial skills, they will learn how to plan, manage and co-operate to achieve the community's goals. The scheme could increase participatory democracy among the urban poor in housing developments in low-income segments as they would no longer be an under-represented group always asking for and receiving help from the government. They could also be more accepted as legitimate and integral parts of the city as they would be empowered to contribute to housing development.

Table 12.2 Performance of the Baan Mankong scheme by region, 2004–2009

Region	*Number of projects*	*Number of communities*	*Number of households*	*Budget approved (million Baht)*
BMR	266	376	33,428	2,679.26
Centre	120	164	12,047	903.33
Northeast	176	303	14,529	895.7
South	101	234	11,117	627.33
North	82	242	9,080	405
Total	745	1,319	80,201	5,510.62

Source: Adapted from CODI (2009)

Conclusion

In 1972, efforts to solve housing problems in Thailand, especially in Bangkok, began with the statutory establishment of the NHA, which started the era of formulating and implementing a concrete housing policy in the country. Nonetheless, housing policy changes have occurred alongside circumstances that have limited the NHA's operation. The objective of mitigating the housing problems of Bangkok's poor through the construction and provision of formal public housing was withdrawn in the early period of its operation because of financial constraints. While the NHA kept building public housing for middle- and high-income groups to subsidise low-income units, at the same time, it had to turn to a minimal housing approach emphasising such projects as slum-upgrading to cater to the urban poor, who could not afford to purchase high-priced formal public housing units. The demarcation of polarised housing segments for the urban well-off and the poor has become manifest when housing development by the private sector has thrived and dominated Bangkok's housing sector, eclipsing public housing activities under the NHA's operation. High economic growth, together with active financial activities, contributed to the robustness and domination of private-sector housing, which therefore became the main housing provision.

However, thriving housing development in the private sector turned speculative until the economy and housing sector collapsed in 1997, damaging the robustness of housing activities in Bangkok. After the crisis, healing measures in both demand and supply side were instigated, mostly in the private sector, to revive the shrinking housing industry. While the housing problems of Bangkok's poor have not been solved under a housing market dominated by private developers, the government recaptured the role of the NHA and public housing initiatives under "A Million Housing Units" at the beginning of the 2000s. These initiatives include the Baan Eua Arthorn low-cost formal housing and the Baan Mankong slum-upgrading scheme. The post-crisis revival of housing activities in Bangkok has nonetheless produced different results in housing wellbeing, with varied social implications across both the middle- and high-income and low-income segments.

The active development of private housing since the 1980s has enhanced housing opportunities for the growing number of middle- and high-income people in Bangkok. With greater access to housing loans from the GHB and commercial banks increasing their purchasing power, apparently, they enjoy the most leeway in choosing where to live. However, their choice of location is constrained by income. The inner areas of Bangkok have become gentrified, making way for a mass transit system and costly private housing projects (Moore, 2015). The high-income group is more capable of living in inner-city areas, where land and house prices are expensive, while the middle-income earners can afford cheaper housing units on the fringes. Their income level constrains not only living locations but also housing products. As detached houses have become expensive, there have been developments of cheaper townhouses and condominiums to serve the needs of the middle-income group. The decrease in household size is also a social phenomenon restricting the housing options of many Bangkok families. Therefore, there has been a growing number of more affordable condominium projects to serve smaller households. However, as many developers compete in the private housing market, building housing units of better quality with strong security systems is used as a selling point to attract customers, contributing to better living conditions. As Bangkok is notorious for traffic congestion, housing projects, especially condominiums, built adjacent to the mass transit system are more sellable and attractive to buyers desiring a short commute from home to work. A remarkable social innovation for common living in the middle- and high-income groups in the private-sector built housing projects is the Juristic Person system. It is compulsory for condominium projects but still optional for housing estates such as townhouses and detached houses. With the primary aim to protect the buyers' rights after the sale of units, the Juristic Person system is also a mechanism to manage the use and maintenance of common property, including the enforcement of common living regulations, to settle problems among unit owners through formal resolution by means of voting in general meetings or a committee.

Despite more affordable housing built by private developers because of increased income along with good economic performance, the opportunity to own homes still belongs to people with decent incomes, mostly the middle- and high-income groups, ruling out people earning just daily wages. Therefore, the government retook the role of providing public housing during the early 2000s through the existing national housing agency, the NHA, which was eclipsed by the private sector, to solve the housing problems of the urban poor. The Baan Eua Arthorn low-income public housing and the Baan Mankong slum-upgrading scheme were initiated with different objectives. The Baan Eua Arthorn project provided low-income people with formal housing units such as walk-up flats and small detached houses based on a lease-purchase scheme. However, this initiative repeated the unsuccessful public housing scheme developed by the NHA since its establishment. Unsustainability was reflected in the high costs of construction that pushed the selling prices too high for the poor to afford. The project was therefore halted, and it did not meet its prime objective of serving the housing needs of the have-nots.

The most promising pro-poor housing option lies in the Baan Mankong scheme, which includes community participation. The project emphasises community empowerment in housing development through community members' own efforts. With secure tenure on occupied land, they can plan, manage, and construct their own houses based on on-site upgrading of the existing dwellings by mutual agreement. Other options such as relocation are also granted as alternatives to suit the individual community's circumstances. Through implementation of the CODI, the government provides the community with loans and technical assistance, and community members pay back the loans with community savings. The implementation is thus communal rather than serving the needs of individual families or dwellers. Advocacy of the Baan Mankong scheme has been continuously sustained and supported by the government. The current government pledged to support informal communities with the Baan Mankong scheme under a 3-year short-term plan and a 10-year master housing plan benefiting mostly the low-income group (*Prachachat Thurakij Online*, 2016). The merit of social development of housing in Bangkok is truly embedded in the Baan Mankong scheme, which explicitly reflects a process of participatory democracy in housing development, which could be found neither in middle- and high-income housing dominated by the private sector nor in formal low-income housing initiatives under the NHA's responsibility.

Notes

1 1 THB = 0.03 USD (as of November 2017).
2 From January 1, 2013, the starting salaries of civil servants with Bachelor's degrees rose to 15,000 Baht per month. However, the rate does not apply to employees in the private sector. A survey revealed that many still earned less than this threshold.

References

Askew, M. (2002). *Bangkok: Place, practice and representation*. London: Routledge.

Bangkok Metropolitan Administration (BMA). (2010). *Statistical summary 2009*. Bangkok: Bangkok Metropolitan Administration.

Bank of Thailand. (2017). *Real estate business indicators*. Available at http://www2.bot.or.th/statistics/ReportPage.aspx?reportID=102&language=th, accessed on October 19, 2017.

Bhatkal, T. & Lucci, P. (2015). *Community-driven development in the slums: Thailand's experience*. London: Overseas Development Institute.

Boonyabancha, S. (2005). Baan Mankong: Going to scale with 'slum' and squatter upgrading in Thailand. *Environment and Urbanization*, *17*(1), 21–46.

Boonyabancha, S. (2009). Land for housing the poor – by the poor: Experiences from the Baan Mankong nationwide slum upgrading programme in Thailand. *Environment and Urbanization*, *21*(2), 306–323.

Buracom, P. (1987). *Limits of state interventions: The political economy of housing policies in Thailand*. Doctoral thesis submitted to Northwestern University.

CB Richard Ellis. (2012). *Bangkok condominium and housing market report*. Bangkok: CB Richard Ellis.

Chairungrojsakul, P. (2009). *Legal problems concerning on land transaction: a case study of the power and duties of juristic person of housing management*. Master's thesis submitted to Sripatum University, Chonburi Campus.

Chiu, H. L. (1984). Four decades of housing policy in Thailand, *Habitat International*, 8(2), 31–42.

Chulasai, B. (1983). *L'evolution des lodgements urbains à Bangkok – Thaïlande*. *[The evolution of urban housing in Bangkok, Thailand]*. Doctoral thesis submitted to Unité Pedagogique D'Architecture No. 1, Paris.

Community Organizations Development Institute (CODI). (2009). *Baan Mankong Project Centre*. Available at www.codi.or.th, accessed on October 3, 2016.

Community Organizations Development Institute (CODI). (2012). *Baan Mankong: Thailand's city-wide, community-driven slum upgrading and community housing development at national scale*. Bangkok: Community Organizations Development Institute.

Fernquest, J. (2013, 18 November). Starting salaries: Government vs. companies. *Bangkok Post*. Available at www.bangkokpost.com, accessed on 27 October, 2017.

Foo, T. S. (1992a). Private sector low-cost housing. In K. S. Yap (Ed.), *Low-income housing in Bangkok: A review of some housing sub-markets*. Bangkok: Asian Institute of Technology, pp. 99–114.

Foo, T. S. (1992b). The provision of low-cost housing by private developers in Bangkok, 1987–89: The result of an efficient market? *Urban Studies*, *29*(7), 1137–1146.

Hara, Y., Hiramatsu, A., Honda, R., Sekiyama, M., & Matsuda, H. (2010). Mixed land-use planning on the periphery of large Asian cities: The case of Nonthaburi province, Thailand. *Sustainability Science*, *5*(2), 237–248.

Huwanan, O., Radomsuk, P. & Iamnoi, A. (1991). *The attitude of low, middle, and high income people towards the Lease City housing project*. Bangkok: National Housing Authority.

Japan International Cooperation Agency (JICA). (2013). *Data collection survey on housing sector in Thailand: Final Report*. Tokyo: International Development Center of Japan, Inc.

Kasikorn Bank. (2017). *Buying home guidebook*. Bangkok: Kasikorn Bank.

Khongpaen, S. (2003). *'Baan Mankong' for enhancing secured tenure for the urban poor in slum communities*. Bangkok: Community Organizations Development Institute.

Kirinpanu, S. (1993). Housing development in the Bangkok Metropolitan Region. *Paper presented at the International Workshop on Metropolitan/Regional Development*, 11–14 October. Shanghai: Tongji University.

Kramer, T. (2011). *Gated communities and social relations in Bangkok*. Bachelor's thesis submitted to Universitiet van Amsterdam.

Kritayanavaj, B. (2002). Financing affordable homeownership in Thailand: Roles of the Government Housing Bank since the economic crisis (1997–2002). *Housing Finance International*, *17*(2), 15–25.

Krungthep Thurakij Online. (2009, 9 October). Prime minister to support CODI with 1.5 billion for Baan Mankong. Available at www.bangkokbiznews.com, accessed on 16 October, 2017.

Leightner, J. E. (1999). Globalization and Thailand's financial crisis. *Journal of Economic Issues*, *33*(2), 367–373.

Ministry of Social Development and Human Security. (2017). *Summary report of intensive government policy implementation* (12 September 2016–30 April 2017). Bangkok: Ministry of Social Development and Human Security.

Moore, R. D. (2015). Gentrification and displacement: The impacts of mass transit in Bangkok. *Journal of Urban Policy and Research*, *33*(4), 472–489.

National Housing Authority (NHA). (1986). *Introduction of the National Housing Authority*. Bangkok: National Housing Authority.

National Housing Authority (NHA). (2008). *Progress report of the Baan Eua Arthorn*. Bangkok: National Housing Authority.

National Statistical Office (NSO). (1980, 1990, 2010). *The population and housing census, Bangkok Metropolis*. Bangkok: National Statistical Office.

Phongpaichit, P. & Baker, C. (2002). *Thailand: Economy and politics*, 2nd Edition. New York, NY: Oxford University Press.

Poapongsakorn, N. (2013, 13 July). Families are earning more, but living apart. *Bangkok Post*. Available at www.bangkokpost.com, accessed on 27 October, 2017.

Pornchokchai, S. (2002). Bangkok housing market's booms and busts, what do we learn? *Paper presented at the Pacific Rim Real Estate Society*, 21–23 January. Christchurch, New Zealand.

Pornchokchai, S. (2010). Bangkok housing prices: 1991–2010. *Asia-Pacific Housing Journal, 12*(4), 58–62.

Posriprasert, P. & Usavagovitwong, N. (2006). Communities' environment improvement network: Strategy and process toward sustainable urban poor housing development. *Journal of Architectural/Planning Research and Studies, 4*, 54–70.

Prachachat Thurakij Online. (2016, 12 January). Cabinet approved a 10-year low-income housing planned to firstly realised in Ladphrao-Din Daeng Area. Available at www.prachachat.net, accessed on 22 October, 2017.

Pugh, C. (1994). Housing policy development in developing countries: The World Bank and internationalization, 1972–1993. *Cities, 11*(3), 159–180.

Real Estate Information Center. (2010). *REIC Research Report*. Available at www.reic.or.th/reicnews/reicnews_index.asp?p=2&s=15&cmbNewsType=5&keyword=, accessed on 19 November, 2016.

Rujopakarn, W. (2003). Bangkok transport system development: What went wrong? *Journal of the Eastern Asia Society for Transportation Studies, 5*, 3302–3315.

Supawittayanan, A. (2003). Baan Eau Arthorn evaluation project in affordable low income people. M.A. *Research Paper*, National Institute of Development Administration.

Tanaboriboon, Y. (1997). Bangkok traffic. *IATSS Research, 17*(1), 14–23.

Tanaphooma, W. & Bart, D. (2015). An overview of public housing characteristics and living satisfactions: Old and new public housing project in Bangkok. *Procedia Environmental Sciences, 28*, 689–697.

Tanphiphat, S. (1983). Recent trends in low-income housing development in Thailand. In Y. M. Yeung (Ed.), *A place to live: More effective low-cost housing in Asia*. Ottawa: International Development Research Centre, pp. 103–120.

Tanphiphat, S. (1993). Housing finance in Thailand: Prospects for the 1990s. *Paper presented to the International Symposium on Urban Housing*, 24–27 November, Bangkok: Chulalongkorn University.

United Nations Children's Fund (UNICEF). (2009). *Thailand statistics*. Available at www.unicef.org/infobycountry/Thailand_statistics.html#67, accessed on 23 November, 2016.

Vanitchvatana, S. (2007). Thailand real estate market cycles: Case study of 1997 economic crisis. *GH Bank Housing Journal, 1*(1), 38–47.

Verapreyagura, P. (2006). Housing standard certification by the National Housing Authority. *Journal of Architectural Planning Research and Studies*, 4(2), 189–204.

Yap, K. S. (1996). Low-income housing in a rapidly expanding urban economy: Bangkok 1985–1994. *Third World Planning Review, 18*(3), 307–323.

Yap, K. S. (2014). Housing as a social welfare issue in Thailand. In J. Doling & R. Ronald (Eds.), *Housing East Asia: Socioeconomic and demographic challenges*. London: Palgrave Macmillan, pp. 227–246.

Yap, K. S. & Kirinpanu, S. (2000). Once only the sky was the limit: Bangkok's housing boom and the financial crisis in Thailand. *Housing Studies, 15*(1), 11–27.

13 Housing policy and social development in Indonesia

Connie Susilawati

Introduction

Global economic trends and local economic policies are the major drivers in the housing market trends and the changes in housing policies in Indonesia. The Asian financial crisis in 1998 caused exchange rates to drop only 30 per cent from the value prior to the outbreak of the economic crisis. The closure of 67 privately owned banks, increased interest rates, negative GDP growth (–13.6 per cent) and very high inflation rates (65 per cent) caused many property developments to be discontinued.

In addition, major transformations and democratisation of the Indonesian government influenced the changes of the Housing and Settlement Areas Act no. 1/2011. Challenges identified in the implementation of socially integrated housing policy include producing more affordable houses and mixed-income residential areas. The majority of the low-cost housing developments are high-density, low-rise houses or very small lots of attached houses. Private developers build only a small number of low-cost housing units; the government also builds a limited amount of public housing, and the rest of the community builds an unplanned informal housing sector to meet housing needs. The government offers some infrastructural funding to improve the liveability conditions of the informal housing neighbourhoods.

In 1974, the central government established three key institutions to implement the housing program: the National Housing Policy Agency (Badan Kebijakan Perumahan Nasional – BKPN), the National Housing and Urban Corporation (Perumahan Nasional-Perumnas) and the State Savings Bank (Bank Tabungan Negara – BTN). The central government of Indonesia has offered a housing subsidy in the form of lower interest rates for low-income borrowers to purchase subsidised low-income housing. The Indonesian government introduced the housing saving program (Tabungan Perumahan Rakyat – Tapera) in 2016, which allows low-income people not just to access subsidised loans but also to upgrade their self-built housing or make additional down payments to enter the mortgage market.

Due to land scarcity in the existing urban area, the government and some developers offer low-cost, high-rise apartments as an alternative housing solution. In addition, the high cost to build high-rise apartments, limited government tax incentives, and low rent fees have made the low-cost apartments uneconomical for private investment. Therefore, the government is also exploring the potential of public-private partnerships to deliver low-cost apartments for low-income employees.

This chapter discusses the housing policies and social development in Indonesia. The first section provides background on the global economic trends and impacts on Indonesian economic policies, followed by the Indonesian social and political conditions. After deliberating the housing policy challenges and changes, the last two sections discuss the impacts of economic trends on housing and the influence on social development in Indonesia.

Global economic trends and Indonesian economic policies

Global economic trends and local economic policies are major drivers in the housing market trends and changes in Indonesia's housing policies. The Asian financial crisis in 1998 caused exchange rates to drop 30 per cent from the value prior to the economic crisis. The closure of 67 privately owned banks, 11 of which were taken over by the government, added to the then-existing 7 state-owned banks (Hawkins & Mihaljek, 2001). The increase of interest rates (over 50 per cent), negative GDP growth (–13.6 per cent) and very high inflation rates (65 per cent) have caused many property developments to discontinue and property developers to go bankrupt. Most development companies rely on either foreign or local bank loans to finance their projects with land as the main collateral of the credit. The unhedged and offshore bad debts caused by many firms could not meet the repayment commitment due to the high demand for US dollars, which caused Indonesia's currency to fall significantly (Nasution, 1998). Unfortunately, the weak financial system and fundamental banking system brought the economy to a crisis. The banks over-extended themselves by providing loans to businesses and individuals. High value credits became bad debts, which eventually caused the crash of the financial market (Winarso & Firman, 2002).

Many companies have been unable to continue their property development. Before the economic crisis, land development had been one of the primary investment sectors in Indonesia (Winarso & Firman, 2002). Some developers targeted a small minority of the richest to produce luxurious housing because it is more profitable than low-income housing. Therefore, the lack of supply of low-income housing created a further gap between the housing market supply and the housing backlog demand of low-income houses.

The government of Indonesia provided an interest rate subsidy as the main economic policy to support low-income households in owning their first homes. Indonesia's first home ownership mortgage credit program, offered by the National Saving Bank (BTN), was introduced on 10 December, 1976. The government policy to subsidise the BTN and Bank Papan Sejahtera keeps their housing loan interest rates low. The supply of simple houses has been supported by low interest rates of the BTN, which for home loans were 9 to 15 per cent, about half of those of commercial banks, which reached 24 per cent on average in April 1989 (Struyk, Hoffman & Katsura, 1990).

Indonesian social and political conditions

Indonesia, a large archipelago, consists of 17,508 islands, but only 6,000 islands are inhabitable (see Figure 13.1). Indonesia is the world's fifth most populated country, with a population estimate of 258,316,051 in July 2016 (The World Factbook, 2016), a majority of whom lived in the five big islands. They speak more than 700 languages, and the formal language is Bahasa Indonesia.

Population and housing

Indonesia is composed of 34 provinces (regional government), and each province has regencies and cities governed by local governments. The most populated area in Indonesia is located on Java Island. Table 13.1 shows that nearly 60 per cent of the Indonesian population live on Java Island.

There are five categories of housing according to Act no. 1/2011: commercial housing, public housing, self-built housing, special housing, and state housing. Commercial housing is built for profit. Public housing is built for low-income people. Self-built housing is housing built by residents individually or collectively with other residents. Special housing is built for a special purpose, such as senior housing, employee housing and student housing. State housing is housing built, owned and operated by the government and allocated to low- to middle-income people (Rukmana, 2015).

The Indonesian government released the "High-rise apartment" Act no. 16/1985, which regulates high-rise commercial and low-cost apartments.

Table 13.1 Projection of population of Indonesia by major provinces

Province	*Capital city*	*2010*	*2015*	*2020*
North Sumatera	Medan	13,028,700	13,937,800	14,703,500
Riau	Pekanbaru	5,574,900	6,344,400	7,128,300
Jakarta special capital region*	Jakarta	9,640,400	10,177,900	10,645,000
West Java*	Bandung	43,227,100	46,709,600	49,935,700
Banten*	Banten	10,688,600	11,955,200	13,160,500
Central Java*	Semarang	32,443,900	33,774,100	34,940,100
Yogyakarta*	Yogyakarta	3,467,500	3,679,200	3,882,300
East Java*	Surabaya	37,565,800	38,847,600	39,886,300
South Kalimantan	Banjarmasin	3,642,600	3,989,800	4,304,000
South Sulawesi	Makassar	8,060,400	8,520,300	8,928,000
INDONESIA		238,518,800	255,461,700	271,066,400

Source: Adapted from Badan Pusat Statistik (2014).

Note
* Provinces located in Java Island.

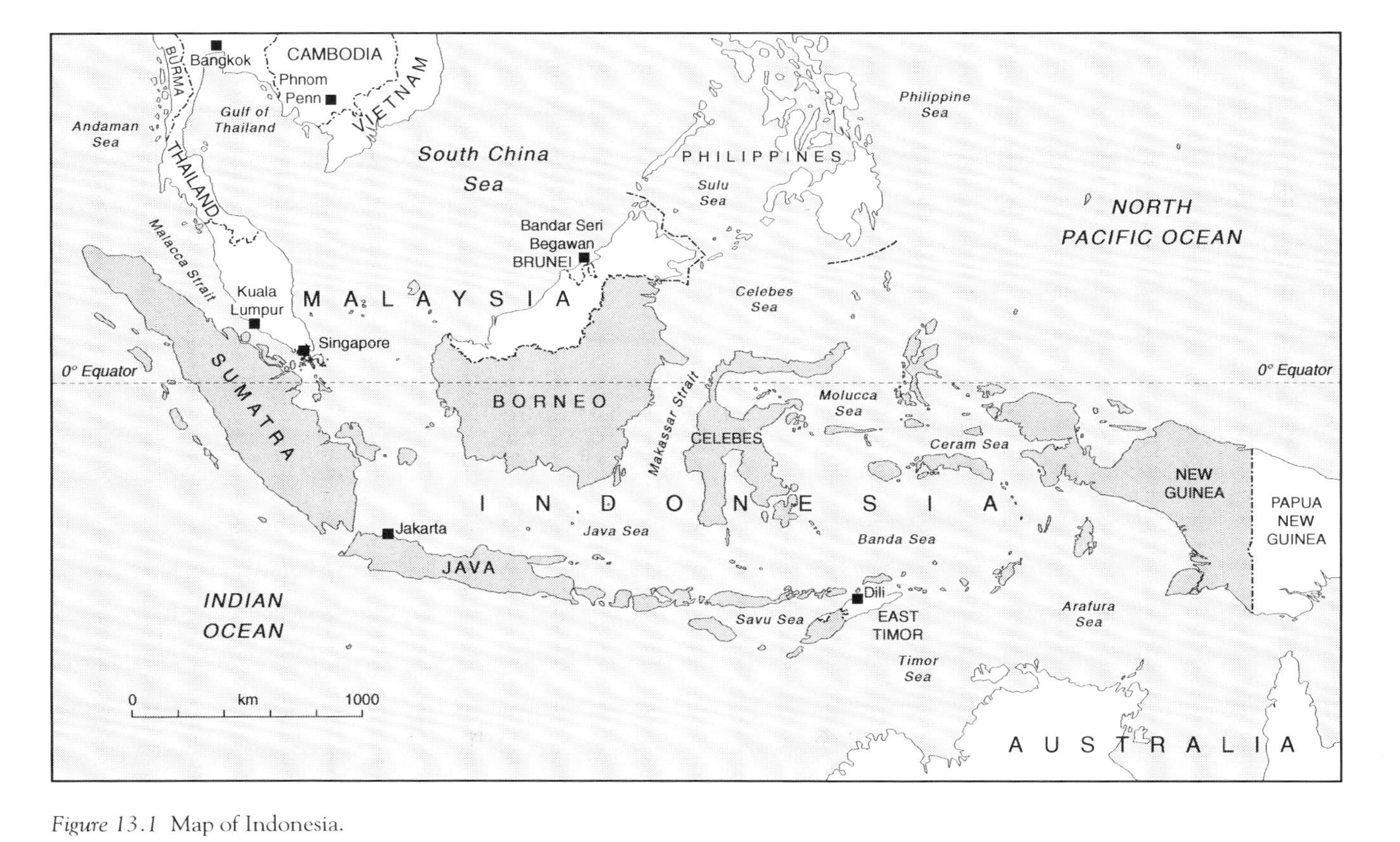

Figure 13.1 Map of Indonesia.

Source: The World Factbook (2016).

The Act regulates the development, ownership, mortgages and occupancy management to respond to the scarcity of land for landed houses in the area near major city centres.

Decentralisation on Indonesian local planning

In the earlier period of independence (1945–1965), the government of Indonesia never endowed local governments with local autonomy. This situation has been changed substantially by Act no. 5/1974, Act no. 22/1999 and Act no. 32/2004, by which decentralisation is interpreted as the transfer of governmental authorities, development functions and affairs from the central government to local governments (Aswad, 2013). In addition to decentralisation, the laws also mentioned deconcentration, which is the delegation of governmental authorities from the central government to governors as its representatives and/or its vertical agencies in particular regions.

The main trigger for transferring the authority to local governments was the economic crisis in 1998 (Wasistiono, 2010). The failure of the central government put Indonesia at risk of disintegration, and the strategic step of implementing decentralisation was used to resolve the crisis.

It explicitly states that decentralisation aims at accelerating the realisation of people's welfare through improving services, empowerment and public participation in parallel with the principles of democracy, equality and justice. Darmawan (2007) suggested that having local planning take charge of local-oriented development and collaboration between local governments and other stakeholders in the development processes is very important.

Role of government on housing policy framework

Following decentralisation, responsibility for housing was also transferred to provincial and local governments. Hoek-Smit (2002) suggested that the outcomes of decentralisation for the housing sector were not clear. The local governments have not had the capacity to plan, finance and deal with local housing markets. The central government (the Housing Ministry) develops the legal and policy framework for housing and the National Housing Policy; sets guidelines, standards and norms for housing; and compiles and maintains a housing data bank and information system to provide technical support and training programs for local and regional government officers. The central government continues to provide subsidies for low-income people so they can afford to buy their first homes.

In addition, the central government plays a supporting role by providing stimulus assistance for self-help housing development (Wilmar, 2015). The central government may intervene in the housing market through financial regulation, fiscal policy, housing policy, investment policy, land policy, planning policy and standards. The local government overlooks the spatial planning and permit systems (UN Habitat, 2008, p. 89).

The regional government (provincial government) will need to coordinate and integrate the regional housing and infrastructure programs (Hoek-Smit, 2002). It must also assess and monitor the progress of local government housing programs funded by the National Housing Agencies. The regional government provides emergency housing in conjunction with the central government. The local government enables the local housing development planning and delivery by private developers, NGOs and CBOs in their local areas.

Political power influenced housing policy

Political power has influenced the housing market in Indonesia. The housing market was distorted in the 1990s, especially in Jakarta and its surroundings (Jakarta, Bogor, Tangerang and Bekasi-Jabotabek), because the developers in Jabotabek are interconnected through family relationships and related to the ex-First Family of Indonesia (Winarso & Firman, 2002). Unfortunately, the relationship has been used to influence any policy and regulation related to land development in the area. Some of the development companies entered the property sector without sufficient financial and technical qualifications and experience (Kemalawarta, 1998). Around 50 per cent of the residential development areas, targeted by small minorities of the rich, have been developed by few developers.

Major transformations and democratisation of the Indonesian government have influenced the changes in the Housing and Settlement Areas Act 4/1992 to Act no. 1/2011. Wilmar (2015) argued that the legislative framework shifts from the stance that housing and settlement developments are for every citizen to participate to become the State's responsibility and to be undertaken by governments. In the new Act, more emphasis is given to low-income housing and slum areas based on the belief that shelter is a basic need for every household. A clear direction and coordination on the formulation of statutory plans on all government levels (central, provincial and local governments) will achieve the government's target of providing affordable housing for all Indonesian citizens and permanent residents.

Housing policy challenges and changes

The housing backlog is estimated at 15 million units (Perumnas, 2015). However, fewer than 400,000 formal homes are developed each year, and this number is not sufficient (Utomo, 2014). The government needs an innovative housing strategy and program to achieve one million houses low-cost apartments to provide houses for all people as part of their basic needs.

The focus of housing policies discussed in this section is supporting the low-income group to achieve their first home ownership. The first group of housing policies supports informal housing, which is built by the community (self-help housing). The second group is the direct provision by the government to provide public housing via Perumnas, a government-initiated organisation.

The third group is the housing policies that enable involvement of the private sector in producing low-cost housing by cross subsidies from the sales of more expensive housing. Finally, housing finance policy is pivotal to ensure that the housing products are affordable and low-income people can access housing through innovative financing solutions.

Self-help housing

Self-help or informal housing is related to the Kampung urban settlement, which is initiated and constructed by occupants and has low quality, limited or no public utilities, and no security of tenure (Tunas & Peresthu, 2010). Low-income people who could not afford to enter the formal housing market were settling in "unused" lots which are not designated for residential use and are near public infrastructure such as railway tracks, canals, rivers, roads, rubbish tips, wasteland, green areas and under bridges.

Indonesia has been implementing three housing improvement policies, including the Kampung Improvement Program (KIP), Community-based Housing Development (Pembangunan Perumahan Bertumpu pada Komunitas-P2BPK) and Self-help Housing Assistance Scheme (Bantuan Stimulan Perumahan Swadaya-BSPS) (Rukmana, 2015). KIP was funded by the World Bank (1969–1982) to improve the living conditions of the existing informal housing. The aim of this program was to improve basic infrastructure and facilities, such as roads and footpaths, drainage, sanitation, solid waste disposal, water supply and building of schools and local health clinics (Tunas & Peresthu, 2010). The KIP program has covered 85,000 ha of slum area in 2,000 locations and helps 36 million people (UNCHS, 1994). However, the World Bank (2003) criticised the programme for focusing on the delivery of limited physical improvement, with no attention to long-term management and integration with the infrastructure network and city development plan. The KIP was reconfigured into the Community Infrastructure Program as part of the Integrated Urban Infrastructure Development Programme (IUIDP). The new KIP included physical, quality of life and economic improvement (Tunas & Peresthu, 2010).

The second program is the P2BPK, which promotes informal and community-based housing. This program involves community participation to provide financial and labour contributions to reduce housing costs (Rukmana, 2015). The third program, BSPS, was launched in 2006 by the Ministry of Public Housing. This program assists low-income households through cash assistance and building materials, including the development of new houses, the improvement of housing quality and the development of public infrastructure and utilities (Ministry of Public Housing, 2011). This assistance is available to low-income households who own the land titles.

Public housing provided by Perumnas, later known as national housing and urban corporation

Wilmar (2015) suggested that the central and local governments should lead the development of low-cost housing through direct provision by Perumnas. Perumnas was launched in 1974 for civil servants and then extended for low-income people with salaries below US$111 per month. Table 13.2 shows the main milestones from 1974 to 2015.

Perumnas is the main government-initiated organisation to build housing for low-income people. Although Perumnas' performance is improving, it definitely could not work alone to meet the housing backlog. Table 13.3 shows the achievements of Perumnas from 2011 to 2015. Table 13.4 illustrates the breakdown of Perumnas' income in 2014 to 2015. All income components in 2015 have increased from the same categories in 2014. Despite the improvement of Perumnas' performance in the post-2008 period, the target of building

Table 13.2 Main milestones of Perumnas, 1974–2015

Year	*Milestones of Perumnas*
1974–1982	Perumnas built simple houses complete with infrastructure for low-income people. A thousand homes were built in Depok, Jakarta, Bekasi and spread out to Cirebon, Semarang, Surabaya, Medan, Padang and Makassar.
1983–1991	Perumnas started building low-cost apartments to help the urban renewal program.
1992–1998	Perumnas built almost 50% of total housing development to support the government program to build 500,000 simple and very simple houses.
1999–2007	Post economic crisis, Perumnas restructured the company's loan and reduced the buying capacity of low-income people.
2008–2009	Perumnas' performance increase achieved target 300% higher than in the previous year. Perumnas led the building of simple rent flats in the 1,000 tower program.
2010–2015	Perumnas became National Housing & Urban Corporation as the main housing providers with a stated target to build 100,000 houses per year.

Source: Perumnas (2016a).

Table 13.3 Operational achievements of Perumnas, 2011–2015

Details	*Unit*	*2015*	*2014*	*2013*	*2012*	*2011*
Sale of house and low-cost apartment	Unit	7,006	7,237	7,686	9,727	9,395
Inhabited low rent flats	Unit	4,868	4,886	5,004	4,953	4,995
Occupancy rate of low rent flats	Percentage	93.3	97.4	98.3	93	96.3
Crude land plot sale Kavling	Plot	196	557	1,038	1,219	563

Source: Adapted from Perumnas (2015, p. 50).

Table 13.4 Perumnas' income breakdown, 2014–2015 (US$)

Description	*2015*	*2014*
House sales	59,143,843	58,825,157
Maintenance lease income	1,880,476	1,760,295
Raw Land Plot Sales	11,419,828	10,232,729
Partnership Income	19,538,068	17,993,925
Low-cost apartment Income	10,391,180	10,222,653
Discounts	–905,474	–
Total Sales	101,467,920	99,023,097

Source: Adapted from Perumnas (2015, p. 150).

100,000 houses per year cannot be reached by it alone (based on the data presented in Table 13.3). Potential collaborative options are discussed in a later section.

Mixed-income housing provided by private developers

A developer has to deliver mixed-income housing to either the same location or a separate development location. The main purpose of Act no. 1/2011, which regulates housing and human settlement and replaces Act. no. 4/1992, is to create harmonious living among people of different social and economic backgrounds. The mixed-income housing will allow cross-subsidies for the procurement of public infrastructure and facilities. It is based on the concept of balanced housing development with the 1:3:6 ratio that was established in 1992. This means that for every 1 unit of the luxury housing category, developers are required to build 3 units of the secondary category and 6 units of the Simple Houses/Very Simple House (RS/RSS) category.

The lack of mixed-income housing is partly due to the lack of supervision and coordination by the government (Sulasmi, 2015). Moreover, developers consider the aspects of profits and demand trends more than community needs and the purchasing power of the community. Developers negotiate with the local government to build public facilities and infrastructure to replace the low-income housing requirements.

In Act no. 1/2011, the government acknowledges that housing is a basic human need and that the nation is responsible for providing the development of affordable, healthy, safe and sustainable housing in Indonesia. Simple houses are landed houses built within the government classification of building-areas and sale costs. The Act and the Housing Ministry Regulation no. 7/2013 require a minimum floor size of 36 square metres and 60 to 200 square metres of land, which follows the standard of a healthy house (9 square metres per person) for a family of four (Ministerial decree on Settlement and Regional Infrastructure no. 403/KPTS/M/2002). The price of a secondary house is at a maximum 6 times the price of a simple house. Finally, luxury housing is sold at a minimum of 6 times the price of a simple house.

Many developers find it very challenging to comply with the 1:2:3 composition ratio established in 2012. Developers need to build two houses for mid-income consumers and three houses for low-income consumers (simple houses) for every luxury unit they build. The land area of simple houses must be more than 25 per cent of the total land area. Tunas and Darmoyono (2014) opined that the reasons private developers had difficulties in building low-income housing were the high value-added tax (10 per cent) and high land prices.

Similar to houses, the construction of mixed-income apartments is regulated by Act no. 20/2011. A minimum of 20 per cent of total units built are targeted for low-cost apartments. They do not have to be in the same building, but in the same area as the commercial apartments. Mixed-income housing is recommended for developers to enable cross-subsidies from commercial apartments to cover the costs of low-cost apartments (Susilawati & Yakobus, 2010).

Housing finance and tax policy

The state saving bank (BTN) provides low-income first homebuyers a subsidised interest rate loan. The current subsidised first home loan program requires only a minimum 1 per cent down payment and low interest rates of 5 per cent for a maximum of a 20-year duration of the loan (Bank Tabungan Negara, 2017). This housing finance policy will enable all Indonesian citizens to access the mortgage market with low down payment requirements.

The government has also offered an innovative housing finance program through the Housing Finance Liquidity Facility (FLPP), down payment assistance, free value-added tax and infrastructure/utilities grants. Since 2010, the government has funded FLPP through the revolving fund of a governmental budget mechanism (Tunas & Darmoyono, 2014). It is a highly subsidized structuring of the total mortgage amount. The government also covers 70 per cent of the credit risk by a mortgage insurance given by Askrindo (a 100-per-cent state-owned insurance company). The mortgage insurance premium is 0.37 per cent, which is included in the final 7.25 per cent interest rate for the duration of a 20-year loan (Utomo, 2014).

The maximum sales price for low-income people to be eligible to receive government subsidy for low deposit and subsidized interest rate varies based on location and years. In 2015, the FLPP was offered only for low-cost apartments with the maximum sales price in the range of US$18,431 to US$41,870 (36 square metre units). The maximum sales price is determined mainly by the building costs, not buying power. There is no linear relationship between the minimum wages and the maximum sales price, so affordability may vary in different regions.

Ministry of Finance regulation no. 113/PMK.03/2014, which is still enforced, states that simple houses that are not larger than 36 square metres are exempted from the 10 per cent value-added tax as long as the price is under the ceiling price determined by the government (see Table 13.5). Such houses are built on sites no smaller than 60 square metres. This benefit is for first-time homebuyers

Table 13.5 Maximum simple landed house prices eligible for subsidised home loans in million rupiahs

No	Location	2014	2015	2016	2017	2018
1	Java (except Jakarta, Bogor, Depok, Tangerang and Bekasi-Jabodetabek)	105	110.5	116.5	123	130
2	Sumatera (except Riau and Bangka/Belitung)	105	110.5	116.5	123	130
3	Kalimantan	115	121	128	135	142
4	Sulawesi	110	116	122.5	129	136
5	Maluku and North of Maluku	120	126.5	133.5	141	148.5
6	Bali and Nusa Tenggara	120	126.5	133.5	141	148.5
7	Papua and West Papua	165	174	183.5	193.5	205
8	Riau and Bangka Belitung	110	116	122.5	129	136
9	Jabodetabek	120	126.5	133.5	141	148.5

Source: Ministry of Finance (2014).

who use a house as principal place of residence and keep it for at least the first five years.

Ministry of Public Works and Housing Decree no. 552/KPTS/M/2016 specifies the maximum income eligibility for the buyers of subsidised housing. The maximum income has increased from US$259 to US$296 per month for buying simple landed houses. Similarly, it increased from US$407 to US$519 for buying low-cost apartments. Further analysis of the ability of low-income people to access subsidised housing is discussed in the following section.

Low-income people need to be able to pay the down payment and also repayments for the duration of loans. In 1986, army personnel started compulsory savings accounts for their own homes after saving for their pensions. Similar arrangements were conducted by air force and also navy personnel. Then, similarly, an arrangement was implemented for civil servants, which was set up by Presidential Decree no. 14/1993 and replaced by Presidential Decree no. 46/1994. The Housing Saving Program for Civil Servants (Badan Pertimbangan Perumahan Pegawai Negeri Sipil-Bapertarum) is supporting civil servants to own healthy and appropriate housing. The scheme supports employees entering the housing finance system by increasing their capacity to pay housing down payments (Bapertarum, 2017). The fund can be used for employees who have worked for a minimum of five years to gain first-time homebuyers' home loans or to improve their existing homes. Additional loans that they access from Bapertarum have a duration of 15 years, with a fixed interest rate of 6 per cent per year.

The latest housing finance initiatives are the establishment of the housing contractual saving scheme and microfinance institutions (Utomo, 2014). Act no. 4/2016 of the Housing Saving Program (Tabungan Pembangunan Rakyat-Tapera) was introduced to collect long-term funding to provide housing financing support to home purchasers. The fund needs to be affordable and long-term to match the long-term housing loan. The existing civil servant house saving program (Bapertarum) will be merged with the Housing Saving Program. The organization will be in full operation on 24 March 2018. The impact of this Act is yet to be evaluated; however, it has great potential based on the success of the predecessor programs for civil servants and defence personnel.

The Housing Saving Program is a new system organised by the government for saving for housing by all employees and self-employed people (Act no. 4/2016). Periodic saving by the members depends on their employers' arrangements to support housing financing, and these savings, including investment outcomes, will be returned including investment outcomes from their saving when their membership ceases.

All workers who receive more than the minimum wage requirements and work more than six months or are self-employed must be members of the Program. The employers must register their workers and the self-employed must register by themselves. Utomo (2014) mentioned that the main barriers for people in Indonesia to access the existing housing finance system are the informal or self-help housing (80 per cent of low-income housing) and informal employment (70 per cent of total workforce).

Similar to the pension fund, membership will cease when members are no longer working (reach pension age), die or are unemployed for five continuous years, including permanent disability or being fired (Act no. 4/2016). At the end of the membership, members can receive all their savings and the weighted average of their net investment income. The investment income arises mainly from house finance payments. Members can choose conventional investments (bank deposits, central government bonds, local government, securities in housing and human settlement and other safe investment) for their membership payments or follow Islamic investment rules (a government bond is called *sukuk*).

Only low-income members can use the fund to help purchase, build or renovate their first home (Act no. 4/2016). The funding is accessible after 12 months of membership and prioritisation based on the membership length, membership saving, urgency of home ownership and availability of the fund. Members can buy only their first simple home (once). The house can be a landed house, row house or high-rise apartment. Funding can be used also to enter the Rent to Buy House Scheme.

Impact of housing policy changes on housing conditions in Indonesia

The impact of housing policy changes will be discussed in four sub-sections: housing affordability, housing security and tenure, liveability and housing culture. The housing affordability is analysed using the current eligibility requirements.

Housing affordability

The Indonesian government has offered two housing programs for low-income people, simple landed housing (Table 13.5) and low-cost apartments. Housing "affordability" is calculated by keeping the housing monthly repayment at less than 30 per cent of the household income. The calculation of monthly repayment uses the current subsidised first homebuyer loan, which has a down payment of 1 per cent of the housing price and a fixed interest rate of 5 per cent per annum for a maximum duration of 25 years. Thus, the interest rate is 5 per cent divided by 12 months (interest rate per month). The duration of the loan is 25 years or 300 months. The present value is 99 per cent of the house price, as the down payment is only 1 per cent. The house prices are listed in Table 13.5.

People with the minimum wage may not be able to afford the maximum sales price. The new prices are affordable only if the first-time homebuyer receives US$296 per month. The maximum eligible income to receive a subsidised first-time homebuyer loan is US$296 per month. With US$296, low-income people pay less than US$89 per month (30 per cent of their monthly income) if they receive subsidised home loans.

The situation is worsened for low-cost apartments. In all areas, they cannot afford to pay the monthly repayment if they receive minimum wages. The required repayment varies between 62 to 123 per cent of their minimum wages. The existing prices for low-cost apartments (Table 13.5) can be affordable for low-income people who receive wages of US$519 per month (maximum eligibility to receive subsidised loan), except for residents of Papua and West Papua. They are required to pay around 30 per cent of their monthly income with the subsidised fixed interest rates of 5 per cent for 25 years.

Housing security and tenure

Hoek-Smit (2002) stated that government policies have emphasized home ownership, but rental housing will continue to be vital to housing market operations and labour mobility. Most of the rental stock is informal housing provided by the private sector, including room rentals and house rentals. Rent is an informal transaction and is not regulated by the government. However, property investors do not rely on rental income as a return on their investment. The high capital gain, which is the main motivation for property investors, unfortunately pushes the housing prices higher.

Indonesia's government has a relatively small proportion of government-assisted and employer-rental housing. Perumnas also built low-cost rental flats (Rusunawa) as mentioned in Table 13.3. In addition, provincial and local governments also built low-rent flats, which will be discussed in a later section.

In some areas, the local government provides leaseholds to enable people to build their homes and live in them. A home can be transferred, and if the inhabitants stay long enough (more than 20 years), they can purchase the land from the local government. In Surabaya, the leasehold title is called a "green letter" (*surat hijau*). Low-income people do not have to purchase the land, but must pay for yearly leases from the local government. The leasehold is the only way for the inhabitants to afford their homes, by gradually building houses when they can (self-help housing).

Firman (2004) mentioned that non-title ownership land belongs to the informal land market, and this land is more difficult to transfer and cannot be used as collateral to apply for loans from banks or other financial institution. Three categories of land tenure are found in low-income settlements: formal land tenure, semi-formal tenure and informal tenure (Reerink & van Gelder, 2010). Land titling has been perceived as tenure security based on the 1960 Basic Agrarian Law. The semi-formal tenure is a native title system (native ownership), which can be converted from "old" colonial native ownership rights to the formal tenure. The informal tenure can be formalised by granting the "new" rights of state land to its occupants.

President Joko Widodo instituted a fast and free process of land certification of formal and semi-formal tenure land by the land rights holders. The cost of certification will be covered by the government's budget and local governments budgets and be part of the corporate social responsibility of large development companies (Pitoko, 2016).

On average, the land administration requires two weeks to process an application, and the period depends on the completeness of the documents, such as the proof of land transfer or proof of semi-formal ownership from the head of a village. Indonesian citizens can have a freehold title; however, foreigners (Indonesian permanent residents) can only have a right to use or right to build for a period of 30 years, which can be extended for 20 years, and renewed for another 30 years (Government Regulation no. 103/2015). However, the rights have to be transferred within one year when the foreigners no longer live in Indonesia.

For low-income people, the stamp duty (BPHTB is Bea Perolehan Hak atas Tanah) cost can hinder the transfer or registration of a land title. After the land area has been confirmed by a surveyor, the applicant can pay a duty of 5 per cent of the tax value of the land. Then, the registration of title and issue of certificate or land title will cost more. It might take six months to release the land title if all the requirements are satisfied.

Liveability

The liveability issues occur mainly in the informal housing sector, where inhabitants build their own homes with limited services. Some houses have not had proper sanitation, as 28 per cent of urban Indonesians do not have access to improved sanitation facilities and 13 per cent (18 million) still practice open defecation (World Bank, 2016). In addition, one-third have access to public utility companies and only 42 per cent of urban households have access to a water supply network.

The government offers some infrastructure funding to improve the liveability conditions of the informal housing neighbourhoods. The housing improvement program for low-income people is not about building a new roof, but it has a comprehensive community-based program. The Slum Alleviation Policy and Action Plan (SAPOLA) is an initiative at the national level to formulate policies and programs on slum alleviation in Indonesia (Utomo, 2014; World Bank, 2016).

The Social Ministry provided funding of US$741 (US$667 for material and US$74 for labour costs) to help home improvement. Some houses are not suitable for people to live in. Kementrian Sosial (2011) listed the guidelines to improve housing with poor conditions (inhabitable houses). The houses that do not meet the health, security and social requirements will need to be upgraded to meet habitable standards:

1. Not built of permanent materials and/or damaged houses
2. Non-durable building materials for wall and roof, such as wood, grass, bamboo
3. Broken wall and roof that put the occupant in danger.
4. Floor from cement or natural soil in broken conditions
5. No bathroom and toilet

Some of the slums or squatters have to be relocated to safer locations. For example, some houses are located at inappropriate locations such as disaster-prone areas, and thus relocation of squatters to multi-story low-income housing is necessary to provide them safe and healthy housing.

Housing culture

The majority of Indonesian housing is informal landed housing built by home owners. The quality of the houses is highly dependent on the owners' ability to afford housing. Some houses are built using temporary building materials until owners can afford to improve the housing conditions. In addition, some houses are extended over time after starting with smaller and over-crowded houses to reduce initial housing costs. When the family grows, the home owner builds additional bedrooms to welcome new members of the family.

The Indonesian Bureau of Statistics (BPS) recognizes that some families use part of their home as a work place, which is categorised as a "household industry" (Dunham, 2009). In addition, the labour force consists of unpaid family workers. Many owners use part of their houses as their workplaces to create the main or additional family income.

Housing is part of a community development programme which also includes people as a factor in housing production. Self-help housing development is not a constituent of the National Housing Strategies (Noor Sidin, 2002). As a result, self-help housing or informal housing development in the inner city areas is overcrowded and associated with poverty and slums.

In contrast, formal housing development projects are conducted mainly by private developers and are located in the fringe areas or outside the established city development. Many large developments on the scale of new city developments are located farther out from the city centre. Although some employment is generated in the new city, the residents might not be able to shift their employment closer to their new homes. Therefore, such development has further burdened the existing road and transport infrastructure to the city centre.

Due to land scarcity and high travel costs and time, the most effective housing solution to optimise the available land is to build taller buildings (Susilawati & Yakobus, 2010). The government can deliver both social and hard infrastructure more cost-effectively. The mass public transportation and road infrastructure can be justified as being more cost-effective because of the increased demand in relatively smaller but more populated land areas. In addition, vertical housing will provide an increased opportunity for open space with the same density.

Prior to 2004, in East of Java, only 8 low-cost apartment projects providing 2,429 units were built in Surabaya. In 1975, the first low-cost apartment was built in a well-located area, Urip Sumoharjo, at the heart of commercial Central Business District. Unfortunately, the other apartments have not been built in the desired location. Thus, the take-up rate was very slow when new apartments were first built and is gradually getting popular as the surrounding area gets developed. Thus, Perumnas and the local government in Surabaya could not pay operational expenses from rental income. Many cases of low occupancy and bad debts have caused the government to provide subsidies to operate low-cost apartments, especially for the earliest projects.

Table 13.6 shows the number of public flats in three adjacent cities–Surabaya, Sidoarjo and Gresik–as recorded by the East Java provincial government (2014). Only five low-cost apartment projects in Gresik provide 579 units, each of 21 and 24 square metres. More than 57 per cent of the units in Gresik are built for student accommodation (tertiary or religious-based boarding schools). In Sidoarjo, the stock managed by both the local government and private organisations is four times that in Gresik. Finally, Surabaya offers ten times the units provided in Gresik, and their sizes range from 18 to 54 square metres.

Housing policy and social development in Indonesia

The One Million Houses Program is a collaborative effort of the central government, local government, private developers and community for low-income people earning between US$185 to US$296 per month. The government has provided three initiatives to support the One Million Houses Program:

1. increase purchasing power by reducing the down payment obligation to 1 per cent of the sales price and a direct subsidy to low-income people;
2. supply infrastructure and utility to reduce the home sales price for low-income people; and

Table 13.6 Number of public flats in Surabaya, Sidoarjo and Gresik

Sum of total units	*Management organisations*				*Grand total*
Row labels	*Perumnas*	*Local government*	*Provincial government*	*Private*	
Gresik	–	579	–	–	579
National budget Housing	–	96	–	–	96
National budget Public work	–	483	–	–	483
Sidoarjo	–	1,722	–	409	2,131
National budget Housing	–	156	–	–	156
National budget Public work	–	1,566	–	192	1,758
Provincial budget	–	–	–	217	217
Surabaya	656	4,056	796	465	5,973
National budget Housing	–	–	–	186	186
National budget Public work	–	3,456	–	–	3,456
National Housing Development	656	480	–	–	1,136
Proposed provincial government land	–	–	528	–	528
Provincial budget	–	120	268	279	667
Grand total	656	6,357	796	809	8,683

Source: East Java Provincial Government (2014).

3 support the revision of Internal Affairs Ministry Decree no. 32/2010 about building permit guidelines to reduce the fee and processing time.

The program is not only for home ownership but also other tenure types such as rental houses, special houses and self-help houses (Dirjen Penyediaan Perumahan, 2016).

In order to support the One Million Houses production, a strategic government plan was launched to streamline the development approval process (Perumnas, 2016b). The original development approval process comprising 42 steps was streamlined to only 8 steps: local environment permit, master plan approval, land use planning approval, principal approval, location permit, environment agency approval, traffic assessment and site plan approval. In addition, the President instructed the government to fast track the development permit to 14 working days for large developments and 9 days for smaller developments.

Land taxation instruments have not been effectively applied to control land utilisation in the cities (Firman, 2004). Two government tax policies, the Sunset Policy tax (2008) and the Tax Amnesty mechanism (2016), have encouraged tax payers to declare their assets in their annual tax reports. The latter have been designed to improve data and information on land administration.

Utomo (2014) stated that the aim of the Indonesian government, as for other nations, is to support the low-income group. Indonesia has classified three groups as the targets of government supports:

1 The very low-income who cannot afford housing: the government directly provides housing through the collaboration of the local government, which provides land, and the central government, which funds the building of multi-storey rental houses for very poor people, charging rent of only US$7.4 to US$18.5 per month;
2 The low-income who can afford to buy homes: the government provides an interest rate subsidy for loans of low and fixed interest rates for the purchase of low-cost formal housing and subsidy incentives for low-income housing developers to build infrastructure, utilities and public utilities.
3 The self-help home owners who need to improve their living conditions: the government provides incentives for housing re-development (US$815) and for housing improvement (US$482).

According to the Property Guide, Surabaya is the cheapest developer to build simple houses, at the cost of US$222 per square metres. For subsidised houses, the building cost for a 36-square-metres simple house is US$8,001, which leaves only US$1,111 for the land cost to qualify for a maximum of US$9,112 (see Table 13.5). It is impossible to find 60 square metres of land for the price of US$18.5 per square metre anywhere in Surabaya. The building cost is cheaper outside of Surabaya, for example US$148 per square metre (because the minimum wages in Surabaya double that in the East Java province), and it is possible to find land prices of US$63 per square metres in East Java outside Surabaya. Therefore, it is not possible to acquire land within the same local government to meet the mixed income housing requirements (1:2:3). The other option is collaboration with local governments by providing leasehold government land for low-income housing development.

Developers also proposed to the government that the developer should be allowed to build simple houses outside the development but within the same province, especially in Java. However, this proposal has not yet been approved under the current regulations. Private developers argued that the high land prices make it impossible to find feasible locations to build simple houses according to the current regulations about ceiling prices and building/site areas. The director of large developments in Indonesia mentioned that the real estate industry proposed modification of the current requirements for simple landed houses, requiring either minimum size or maximum price, but not both.

Susilawati and Yakobus (2010) suggested five essential incentives to attract more developers to build low-cost apartments:

1 Value-added tax exemption to boost the viability of the project,
2 Assistance on the land acquisition and development permit process to support the viability of projects,
3 Government grants for infrastructure and utility installation to help reduce total development cost,
4 Interest rate subsidy to make loans affordable for low-income households, and
5 Review of eligibility requirements for receiving government subsidy for homebuyers. The maximum income of homebuyers has now increased to US$518.6 per month, and the maximum value of apartments is now more than US$18,520 per unit for 36 square metres.

Table 13.7 summarises the collaboration by each stakeholder in the development of low-income housing from the areas of planning, land procurement, infrastructure development and building houses for low-income people. The new Housing and Settlement Areas Law stipulates the housing subsidies and assistance for low-income residents. The government is mandated by the new Housing and Settlement Areas Act to assist low-income residents by offering tax incentives, occupation permits, land and public utilities provisions and land title registration (Rukmana, 2015).

Land scarcity in the existing urban area has pushed the low-cost development outside the city centres. The government and some developers offer low-cost, high-rise apartments near the city centres as an alternative housing solution. However, the high cost to build high-rise apartments, government tax incentives and low rent fees have made the low-cost apartments uneconomical for private investment. Therefore, the government of Indonesia is also exploring the potential of public-private partnerships to deliver low-cost apartments.

Table 13.7 Collaboration across stakeholders on the delivery of low-income housing

Incentives/ funding	*Land*	*Development approval*	*Infrastructure*	*Low-income housing*
Government	Released service land	Fast track	–	VAT-free, subsidised home loan
Private	–	Fast track and zoning	Infrastructure incentives	VAT-free, subsidised home loan
–	–	–	–	(PPP employee housing)
Low-income people	Free titling	Relocation if necessary	KIP	Microfinance

Source: Adopted from Pitoko (2016), Rukmana (2015) and Utomo (2014).

Public-private partnerships have been suggested by a few researchers as a possible solution to increase the supply of affordable rental apartment or low-cost/simple rental flats. Hoek-Smit (2002) suggested that public-private partnerships could be used to build employee housing. Rachmawati et al. (2015a) suggested that the development of employee housing needs to consider inhabitants' needs and regulations as well as initiate partnerships with private sector employers who want to provide housing for their employees on their land. Furthermore, Rachmawati, Susilawati, Soemitro & Adi (2015b) and Rachmawati, Soemitro, Adi & Susilawati (2016) recommended that the government's concerns should be about site availability, the public decision-making process and soliciting a good consortium to form partnerships, while those of the private sector would be the ability to pay rent on the site, macro-economic conditions and policy, housing finance availability and government support.

The changes in housing policy have focused on the area of formal housing, and they will improve the future housing supply. However, the government has paid limited attention to the existing informal housing that was mainly built by home owners. The improvement of livability of informal housing is managed by the Ministry of Social Development, not the Ministry of Housing and Public Works. An integrated effort across government units and their agencies will be required to help the most vulnerable groups to access livable and healthy housing. In addition, the concentration of affordable housing outside the city centre has put more pressure on the transportation infrastructure and increased transportation costs and time, which affect the living costs of the vulnerable groups. Finally, the latest government policy to assist informal and formal housing through housing finance policy has great potential to extend the access to build or to improve housing conditions through the saving program (Tapera).

Conclusion

Housing is a basic need, and the state has recognised that it has responsibility to ensure that all Indonesian citizens and residents have a safe, secure and healthy living environment. The government of Indonesia has invited everyone to take part in this national housing scheme through Tapera. However, as the initiative was introduced only in 2016, this study will not be able to measure the impact of the latest housing policy on social development in Indonesia.

The saving program plus the government subsidy for low-income people will enable them to access the subsidised first-home program. This pool of money will help provide long-term funding to support low-income home ownership and upgrading existing housing. This study shows that eligible households will be able to afford purchasing their first homes. The exemption of the value-added tax (10 per cent), low down payments and fixed and low interest rates for the duration of the loan have been offered through FLPP and Tapera.

On the supply side, the government has provided support through housing improvement programs for self-help (informal) housing, direct provision by Perumnas and mixed-income housing by private developers. Although informal housing is dominant, occurring in around 80 per cent of total housing in Indonesia, the present changes in housing policy have focused on formal housing. Upgrading current houses to meet healthy housing standards will be one of the government's top priorities, not just in newly built housing. The improvement of liveability of the informal housing and the housing finance assistance and subsidy are managed by different ministries. Some homes are used as workplaces as well, affecting the quality of homes. Therefore, an integrated effort across government levels and their agencies will offer housing solutions and increase productivity and social development for low-income people and vulnerable groups.

Perumnas and private developers who build simple landed houses and low-cost apartments for low-income people under the maximum sales price will receive subsidies to build infrastructure. Although mixed-income housing is a requirement for private development, these inclusionary zones have not been implemented in all developments. Furthermore, this study has illustrated the barriers to the supply of houses and that land is too expensive. Thus, the government's provision of land as a leasehold will reduce the total development cost and allow the delivery of low-cost housing. Public-private partnerships have been suggested to provide employee housing. Finally, collaboration between the government and the private sector is essential to ensure that housing is affordable, especially for low-income people.

References

Aswad, S. (2013). *Local development planning and community empowerment in decentralised Indonesia: The role of local planning in improving self-organising capabilities of local communities in Takalar, Indonesia.* PhD thesis submitted to Queensland University of Technology.

Badan Pusat Statistik. (2014). *Population projection per provincial government, 2010–2035 (in million)*. Available at www.bps.go.id/linkTabelStatis/view/id/1274, accessed on October 11, 2017.

Bank Tabungan Negara. (2017). *BTN subsidised home loan* (in Indonesian). Available at www.btn.co.id/id/content/Produk/Produk-Kredit/Kredit-Perorangan/KPR-Bersubsidi, accessed on October 10, 2017.

Bapertarum. (2017). *Company history* (in Indonesian). Available at www.bapertarum-pns.co.id/tentangkami/sejarah/, accessed on October 11, 2017.

Darmawan, S. (2007). The nature and trends of local development planning in Indonesia (in Indonesian). *JICA Module*. Makassar: JICA.

Dirjen Penyediaan Perumahan. (2016). *Tentang Program Sejuta Rumah [About One Million Home Program]*. Available at http://sejutarumah.id/index.php/halaman/detail/13/tentang-program-sejuta-rumah, accessed on October 9, 2017.

Dunham, S. A. (2009). *Surviving against the odds: Village industry in Indonesia*. Durham and London: Duke University Press.

East Java Provincial Government. (2014). Unpublished. *Number of apartments in East of Java*.

Firman, T. (2004). Major issues in Indonesia's urban land development. *Land Use Policy*, *21*(4), 347–355.

Hawkins, J. & Mihaljek, D. (2001). The banking industry in the emerging market economies: competition, consolidation and systemic stability: An overview. *BIS Papers No. 4*. Available at www.bis.org/publ/bppdf/bispap04a.pdf, accessed on October 11, 2017.

Hoek-Smit, M. C. (2002). *Draft implementing Indonesia's new housing policy: The way forward*. Jakarta: Kimpraswil, Government of Indonesia and the World Bank. Available at www.scribd.com/document/100958996/IMPLEMENTING-INDONESIA-S-NEW-HOUSING-POLICY-THE-WAY-FORWARD-Draft, accessed on October 11, 2017.

Kemalawarta, K. (1998). Property business and economic crisis (in Indonesian). *Properti Indonesia*, *54*, 57.

Kementrian Sosial. (2011). *Pedoman Pelaksanaan Rehabilitasi Sosial Rumah Tidak Layak Huni dan Sarana Prasarana Lingkungan [Guideline Implementation of Social Rehabilitation of Uninhabitable and Environmental infrastructure]*. Jakarta: Kementrian Sosial.

Ministry of Finance. (2014). *Ministry of Finance regulation no. 113/PMK.03/2014*. Available at http://apernas.org/files/produkhukum/20141204114037PMK%20No.113%20Tahun%202014.pdf, accessed on October 9, 2017.

Ministry of Public Housing. (2011). *Ministry of Public Housing Regulation 14/2011*. Available at: http://www.bphn.go.id/data/documents/11pmpera014.pdf, accessed on October 9, 2017.

Nasution, A. (1998). The melt down of the Indonesian economy in 1997–1998: Causes and responses. *Journal of Economics*, *11*(4), 447–482.

Noor Sidin, F. (2002). Indonesia. In J. Doling, A. Razali & L. Dong-Sung (Eds.) *Housing policy systems in South and East Asia*. Houndmills, New York: Palgrave Macmillan, pp. 161–177.

Perumnas. (2015). *Annual report 2015*. Available at www.perumnas.co.id/annual-reports/, accessed on October 10, 2017.

Perumnas. (2016a). *History of Perumnas* (in Indonesian). Available at www.perumnas.co.id/sejarah-perumnas/, accessed on October 12, 2016.

Perumnas. (2016b). This is the Government strategy to fast track the realisation of one million house. Avaliable at www.perumnas.co.id/ini-strategi-pemerintah-percepat-realisasi-sejuta-rumah/, accessed on October 10, 2016.

Pitoko, R. A. (2016, November 18). Important, since 2017 land certification is free (in Indonesian). *Kompas*. Available at http://properti.kompas.com/read/2016/11/18/1700 00721/penting.mulai.2017.sertifikasi.tanah.gratis, accessed on December 12, 2016.

Rachmawati, F., Soemitro, R. A. A., Adi, T. J. W. & Susilawati, C. (2015a). Low-cost apartment program implementation in Surabaya metropolitan area. *Procedia Engineering, 125*, 75–82.

Rachmawati, F., Soemitro, R. A. A., Adi, T. J. W. & Susilawati, C. (2015b). Critical success factor in partnerships between major stakeholders in low-cost apartment development in Surabaya Metropolitan Area. In *Proceedings of the 2nd Makassar International Conference on Civil* Engineering (MICCE 2015), Makassar, Indonesia, August 11–12, 2015.

Rachmawati, F., Susilawati, C, Soemitro, R. A. A. & Adi, T. J. W. (2016). Major stakeholder different perspective concerning factors contributing to successful partnerships in low-cost apartment development in Surabaya Metropolitan Area in Indonesia. In *22nd Annual Pacific-Rim Real Estate Society Conference*, Sunshine Cost, Queensland, Australia, January 17–20, 2016.

Reerink, G. & van Gelder, J. L. (2010). Land titling, perceived tenure security, and housing consolidation in the kampongs of Bandung, Indonesia. *Habitat International, 34*(1), 78–85.

Rukmana, D. (2015). *The transformation of Indonesian housing policies*. Available at http://indonesiaurbanstudies.blogspot.com.au/2015/09/the-transformation-of-indonesian.html, accessed on December 12, 2016.

Struyk, R. J., Hoffman, M. L. & Katsura, H. M. (1990). *The market for shelter in Indonesian cities*. Washington, D.C.: The Urban Institute Press.

Sulasmi, S. (2015). *Study of policy implementation on housing and settlement development with a balanced environment dwelling*. Master thesis submitted to Institut Teknologi Bandung. Available at http://digilib.itb.ac.id/gdl.php?mod=browse&op=read&id=jbpti tbpp-gdl-srisulasmi-21490, accessed on December 12, 2016.

Susilawati, C. & Yakobus, S. (2010). New affordable strata title housing solutions: A case study in Surabaya, Indonesia. In *Proceedings of 2010 International Conference on Construction and Real Estate Management*, Brisbane, December 1–3, 2010.

The World Factbook. (2016). East & South-East Asia: Indonesia. Available at www.cia.gov/library/publications/the-world-factbook/geos/id.html, accessed on December 12, 2016.

Tunas, D. & Darmoyono, L. (2014). Indonesian housing development amidst socio-economic transformation. In J. Doling & R. Ronald (Eds.) *Housing East Asia: Socio-economic and Demographic Challenges*. Basingstoke: Palgrave Macmillan, pp. 91–115.

Tunas, D. & Peresthu, A. (2010). The self-help housing in Indonesia: The only option for the poor? *Habitat International, 14*, 315–322.

UN Centre for Human Settlements (UNCHS). (1994). *Global report on human settlements 1994*. Nairobi: UNCHS/Habitat.

UN Habitat. (2008). Public intervention in the housing market in Indonesia: Who gets to benefit? In UN Habitat (Ed.). *The role of government in the housing market: the experiences from Asia, the human settlements finance and policies series*. Nairobi: UN-Habitat, pp. 83–108.

Utomo, N. T. (2014). *Affordable housing finance policies in Indonesia.* The World Bank. Washington DC, 28–29 May 2014. Available at http://siteresources.worldbank.org/FINANCIALSECTOR/Resources/Session2_NugrohoTriUtomo.pdf, accessed on October 10, 2017.

Wasistiono, S. (2010). Towards balanced decentralisation (in Indonesian). *Journal Ilmu Politik AIPI No. 21*, 31–53.

Wilmar, S. (2015). Governing housing policies in Indonesia: Challenges and opportunities. In *RC21 International Conference on the Ideal City: Between Myth and Reality. Representations, Policies, Contradictions and Challenges for Tomorrow's Urban Life*, Urbino, Italy, August 27–29, 2015.

Winarso, H. & Firman, T. (2002). Residential land development in Jabotabek, Indonesia: Triggering economic crisis? *Habitat International, 26*, 487–506.

World Bank. (2003). Sustainable development in a dynamic wold: Transforming institutions, growth and quality of life. *World Development Report 2003.* Available at www.rrojasdatabank.info/wdr03/complete.pdf, accessed on January 10, 2017.

World Bank. (2016). *Project Information Document (PID) appraisal stage (Indonesia national slum upgrading project).* Available at http://documents.worldbank.org/curated/en/755311468261274690/pdf/PID-Appraisal-Print-P154782-06-06-2016-1465189135070.pdf, accessed on December 12, 2016.

14 Conclusion

Asian housing policies in the social development contexts

Rebecca L. H. Chiu

The twelve country/city chapters in this book address its central theme from different angles and with different emphases, depending on the housing development trajectories and the major issues confronting the specific communities now. Given this diversity and the individual choices of the narrative if not analytical frameworks, it is inappropriate and impossible to develop a theory on housing policy and social development in Asia in this concluding chapter. However, it is possible to highlight and compare the major features of the nexuses between housing policy and social development in countries/cities of similar economic development levels and growth rates (Table 14.1) to provide a deeper understanding of how the different social settings and their changes have influenced housing policy development, which has subsequently shaped and reshaped the housing wellbeing of the local people specifically and the social

Table 14.1 GDP per capita 2016 and growth rates of major Asian economies, 2013–2017

12 countries[1]	*GDP/capita (current US$)*	*GDP growth rate (%)*				
	2016	*2013*	*2014*	*2015*	*2016*	*2017 (2nd quarter)*
Singapore	52,961	3.9	2.9	2.0	2.0	2.9
Hong Kong	43,681	3.1	2.8	2.4	1.9	3.8
Japan	38,895	2.0	0.3	1.1	1.0	1.4
Korea	27,539	2.9	3.3	2.8	2.8	3.2
Taiwan	22,561	2.2	4.0	0.7	1.5	2.1
Malaysia	9,503	4.7	6.0	5.0	4.2	5.8
China	8,123	7.6	7.5	7.0	6.7	6.9
Thailand	5,908	2.7	0.9	2.9	3.2	3.5
Indonesia	3,570	5.6	5.0	4.9	5.0	5.2
Vietnam	2,186	5.4	6.0	6.7	6.2	6.3
India	1,709	4.5	4.7	7.2	7.6	7.0
Bangladesh	1,359	6.0	6.1	6.6	7.1	7.3

Source: National Statistics, Republic of China (2017), World Bank (2017).

Note

1 Listed in descending order of GDP per capita.

development of a place generally. The comparison also enables the conceptualization of the key relationships between housing policy and social development in Asia to provide pointers for future research.

The Singapore and Hong Kong narratives provide a very interesting comparative pair. In these two of the region's most advanced and globally significant economies, although supporting economic development was a common reason for introducing a massive public housing program, their housing trajectories took opposite tracks. The former strives for home ownership, and the latter targets rental housing; but both have been directly charted and crafted by government, using public resources to meet developmental needs and aspirations in the earlier stage of economic development. It may be argued that the continued undertaking of what are commonly conceptualized as an asset-based welfare policy in Singapore and security-based one in Hong Kong is due to differences in political circumstances and social needs: the building up of a new nation and the political party's quest for sustained power in the former and the survival of a colonial government in the latter. The indigenous government of Hong Kong, established since 1997, has attempted to shift to the Singapore track but has been stifled by unfavourable economic conditions and property cycles.

In facing the similar challenges of an ageing population asking for post-retirement protection, exacerbating housing inequality due to globally induced housing price hikes and incoming migration and changes or potential changes of government leaders, the policy trends tend to converge. Singapore has rekindled public rental housing provision, and Hong Kong has re-pledged subsidized home ownership growth. However, the Singapore government, adopting an asset-based welfare policy, faces a greater dilemma of housing price control whereas the Hong Kong government, adopting a security-based policy led by public rental housing, is restricted by a transient land supply shortage, which seems to be a lesser problem than that challenging the Singapore government. What both have achieved in terms of housing and social wellbeing are satisfactory housing quality, housing security, housing affordability, social stability and post-retirement protection. Singapore's achievement is doubtlessly greater as its housing benefits cover a large majority of the citizens.

Although there is no sign of significant domestically-induced economic stagnancy or externally-driven economic depression in Hong Kong and Singapore, the experience of Japan, the super economy in Asia in the 1970s and 1980s, in tackling housing issues in a post-growth and super-aged contexts rings the alarm bell. Japan's post-war home ownership strategy, which primarily relied on subsidized financial tools supplemented by the government's direct provision of affordable housing, was entrusted with important social and economic missions: building up the middle-income families as a social mainstream and boosting the construction industry to help revive the economy. These missions were achieved, and they brought in housing wellbeing. However, as economic growth has stagnated since the 1990's and the population reached a super-aged level in 2007, the advantages of an asset-based housing policy turned around to exacerbate housing inequality and consequently social polarization, not just in the current generation but in future generations due to the common intergenerational transfer of housing assets in Asian communities. In a post-growth economy in which the government has to minimize housing subsidies, decreasing employment opportunity and wages hamper housing affordability, and falling asset values reduce the post-retirement protective function of home ownership. The outlook for housing wellbeing and the social function of housing is bleak. Perhaps the Japan case demonstrates the longer-term weaknesses of an asset-based housing and welfare policy implemented mainly through a cash subsidy approach. Assuming that state land ownership is mandated or attainable, a land-based subsidy approach such as that of Hong Kong and Singapore can mitigate the impact of market fluctuations on subsidized home ownership. However, it is still onerous on the government to stabilize housing prices to maintain the post-retirement protective function of home ownership, as in the case of Singapore.

The social focus of housing policy in the other three more stable and developed economies in Asia came only in the 2010s. Taiwan's high home ownership rate can be traced back to the full home-owning policy of the 1950s, upheld until 2005. While public housing was built for sale, the major subsidy tool was a mortgage interest subsidy to help build or buy housing in the market. The government also intervened with financial and tax measures when the housing market became over-heated. In a phenomenon driven by the electorates, it was only in 2011 that 'housing quality', 'a dignified housing environment' and 'equal rights to housing' were officially pursued for the first time; the target then began to include providing social housing to the socially and economically vulnerable households. Subsequently, housing justice was officially enshrined in 2014 to constitutionalize housing rights, to provide public rental housing, and to introduce other regulatory and fiscal measures. Further, in 2016, under the new government, social justice and non-government participation in housing provision were officially included. Thus, the social focus of Taiwan's market-driven housing system came relatively late, but once introduced, it moved fast. Similarly, attention to the underprivileged also came late, in 2012 in Korea, with a special emphasis given to re-integrating disadvantaged groups through housing programs. Unlike Taiwan, the Korean government has been heavily intervening in the housing market to stabilize housing prices and control speculation. This effort takes an alternative strategy of increasing supply and controlling housing price by regulatory, legal, planning, and other means.

While the changes in housing policy of the above countries/cities have been gradual, reflecting the path-dependent nature of housing policy, those of China and Vietnam have been abrupt, ushered in by structural changes in the economy. Moving from a centrally planned economy to a market economy but with socialist characteristics, China's housing system was dramatically commodified between 1980 and 2000. By 2010, the private housing market had become dominant, and the ills of a marketized system, such as overheated housing prices and housing inequality, had surfaced simultaneously. The ambitious affordable housing program of 2011 in the main reverted to larger scales of state provision of subsidized housing to target groups. This program aims to address the problems incurred by the market and those of a pro-growth market economy and to achieve the then newly introduced social goal of building a harmonious society. However, the China chapter contends that the main purpose of the housing program is to ensure the accomplishment of the 2013 new urbanization strategy, which underpins the re-orientation of the Chinese economy from export-driven to domestic demand-led. Hence, the latest shift in China's housing policy has continued to be growth-driven, taking a 'productivist' rather than a 'developmentalist' approach, although it includes a social wellbeing slant. Evidence from Shanghai is its targeting of the young and skilled migrants, rather than the low-income groups, to be residents of well-designed public rental housing. Vietnam also underwent a similar economic reform and marketized the housing system to facilitate economic growth. However, both market and public housing in the formal sector is found to be unaffordable for average households, let alone the vulnerable groups. The informal sector is indeed the anchor of the housing supply, and the state has to tolerate the informality, as the formal sector propelled by the state cannot cater to the housing needs of the community. This peculiar state-market-community relationship in housing is indeed also found in prime cities in China such as Beijing and Shenzhen, where 'small property right housing' in the urban fringes prevails. These housing clusters are not necessarily ghettos of low-paid migrant workers but housing enclaves for the better-off rural and urban middle class.

Southeast and South Asian economies except Thailand have also achieved high economic growth rates, not just the post-reform centrally planned economies (Table 14.1). As the more developed economy in the region, Malaysia has long aspired to build a home-owning nation since its independence in 1957. It also emphasizes the provision of adequate, affordable and decent housing for all regardless of income, employment and ethnicity. Various housing programs have been devised and implemented to fulfil these goals, with recent focuses on home ownership opportunities for the lower- and middle-income families and young adults, as well as on housing quality. It may be argued that building a new nation out of a diverse mix necessitates the strong emphases on home ownership, quality and inclusiveness, which these populations aspire to. Similar to the case in Singapore, housing policy in Malaysia is used as a tool to build national pride, social cohesion, equality and living quality.

Such social emphases in housing policy did not get off the ground in Thailand. The first housing policy, which was established in 1972 to solve the housing problems of the low-income groups, failed. Since then formal housing has been dominated by market provision to meet the demands of the higher-income groups. The housing needs of the low-income groups and the poor have been resolved by the cheaper informal housing. The government's financial ability has been the main deterrent to providing formal housing to the low-income groups, even for the recent large-scale project launched in the early 2000s. Nonetheless, economic growth, although impeded by regional and global financial crises at times, has enabled the improved housing standard for the rising higher-income groups as the housing market responds to their demands. Currently, the hope to ameliorate the poor housing conditions of the less affluent seems to lie in the government's providing land security and financial and technical assistance to the residential community for self-upgrading. Similar housing circumstances are found in Indonesia, but the government has attempted to use more diversified approaches to help reduce the housing problems of the low-income groups apart from supporting self-help houses, such as low-interest home loans, housing saving programs and public-private partnership for delivering low-cost apartments in high-rises. Community participation to provide financial and labour contribution is also encouraged to reduce production costs. Improvement of livability in this context refers mainly to meeting basic habitable standards. Community participation facilitated with government assistance is an important government-cum-social approach to deliver better housing wellbeing in Thailand and Indonesia. Nonetheless, stronger government commitment and stamina are found in Indonesia, which is undergoing stronger economic growth.

Despite being economies with the lowest average GDPs per capita as shown in Table 14.1, India and Bangladesh have economic growth rates that are among the highest in the region. In tandem with economic growth and rapid urbanization, India's housing sector has been dramatically marketized along the neo-liberalist approach, resulting in vibrant market growth but simultaneously polarizing housing inequality. As 95 per cent of housing deficit pertains to the low-income and disadvantaged groups, there is a call to reverse the priority of housing policy to facilitate economic growth over improving the housing conditions of the poor. In particular, it is important to channel housing finances to low-income housing and to strengthen institutional and governance structures and the resource base to favour housing development for the poor. More directly, the provision of 'affordable housing' and 'pro-poor housing' is imperative. The problem is found to be greater in Bangladesh, where decent housing is severely lacking for the middle-income group, let alone for lower-income families. Housing produced by the government is usually allocated to government officials although public housing programs aimed at lower-income families are emerging after years of planning on paper. In both India and Bangladesh, lower-income families have been living in slum settlements, mostly with substandard facilities. To these residents, the hope for improving housing wellbeing is still remote, if possible at all, despite the strong economic growth.

At this point, it is useful to return to the definitions of social development cited in the introductory chapter to provide a conceptual lens to understand the nexus between housing policy, housing wellbeing, social change and social development in Asia. Applying the first category, which defines social development as a totality comprising economic, political, social and cultural aspects, and which aspires to improve people's general welfare, it can be concluded that the economic aspect has been a shared and important trigger of a more ambitious and committed housing policy in Asia. The political and social missions are equally important for a new nation or a new government although the promise of housing wellbeing might not be always delivered, depending upon a government's financial capability, land resources, market conditions and economic development level and philosophy. The second set of definitions focusing on producing structural change in the economic and social systems as the core element of social development is pertinent to China and Vietnam. Both countries have undeniably improved housing and social conditions drastically following the marketization of the housing systems. The problem subsequently emerging is polarizing housing and social inequality between the higher- and lower-income groups and between the urban residents and the new migrants from rural areas. The informal housing sector supplying cheaper housing expands accordingly to meet the needs, more so than the government housing programs, which are often formulated under a pro-growth strategy to attract young and skilful workers only.

The third category of definitions focusing on achieving human potential, meeting needs and achieving a satisfactory quality of life are perhaps most apt for the informal settlements in Asia's developing economies. These settlements involve the government's financial and technical assistance as well as active community participation and innovation to jointly develop a satisfactory living environment with minimum means. The organically developed sense of community, indigenously developed innovations and integration with assistance from the formal sector testify to the realization of human potential to achieve more than mere survival in a less-than-desirable living environment. The social quality of life in a housing environment and its importance for building a harmonized society has been upheld by Asian governments. But the social quality of life seems to be a mantra rather than a genuine goal to be realized in most cases. Until this aim is achieved with demonstrable evidence, the amalgamation of social and housing wellbeing in housing policy design remains mere rhetoric.

This book has presented the experiences, trajectories, successes and failures of housing policies in twelve Asian countries/cities and their intricate relationships with social development. More could have been included, and deeper analyses could have been conducted. However, the book represents the first joint efforts of 21 Asian scholars to paint an overall picture of how housing policy and social development interact by telling their respective housing stories. We hope that this volume will attract further collaborations within the region and between Asian and international academia. We believe that through dedicated research, we contribute to the betterment of the housing and social future in Asia and beyond.

References

National Statistics, Republic of China. (2017). *National accounts*. Available at: https://eng.stat.gov.tw/np.asp?ctNode=1539, accessed on 1 December 2017.

World Bank. (2017). *GDP growth*. Available at: https://data.worldbank.org/indicator/NY.GDP.MKTP.KD.ZG, accessed on 1 December 2017.

Index